1st Edition

THE PATENT Drawing BOOK

By **Patent Agent Jack Lo**
and **Patent Attorney David Pressman**

Edited by **Stephanie Harolde** and **Stephen Elias**

NOLO PRESS **BERKELEY**

Your Responsibility When Using a Self-Help Law Book

We've done our best to give you useful and accurate information in this book. But laws and procedures change frequently and are subject to differing interpretations. If you want legal advice backed by a guarantee, see a lawyer. If you use this book, it's your responsibility to make sure that the facts and general advice contained in it are applicable to your situation.

Keeping Up to Date

To keep its books up to date, Nolo Press issues new printings and new editions periodically. New printings reflect minor legal changes and technical corrections. New editions contain major legal changes, major text additions or major reorganizations. To find out if a later printing or edition of any Nolo book is available, call Nolo Press at 510-549-1976 or check the catalog in the *Nolo News*, our quarterly publication.

To stay current, follow the "Update" service in the *Nolo News*. You can get a free one-year subscription by sending us the registration card in the back of the book. In another effort to help you use Nolo's latest materials, we offer a 25% discount off the purchase of the new edition of your Nolo book if you turn in the cover of an earlier edition. (See the "Special Upgrade Offer" in the back of this book.) This book was last revised in **May 1997**.

First Edition	**MAY 1997**
Cover Design	LEO ZUBRITSKY
Book Design	TERRI HEARSH
Illustrations	JACK LO & TERRI HEARSH
Index	SUSAN CORNELL
Proofreading	KIMBERLY A. CLINE
Printing	BERTELSMANN INDUSTRY SERVICES, INC.

Lo, Jack.
    The patent drawing book : how to prepare formal drawings required
by the U.S. Patent Office / by Jack Lo and David Pressman. --
1st ed.
         p.   cm.
    Includes index.
    ISBN 0-87337-378-2
    1. Patents--United States--Drawings.   I. Title.
T223.U3L6 1997
608' .022'1--dc21

                                    97-817
                                      CIP

Acknowledgments

I am most grateful to David Pressman for being a friend and mentor, whose generous support has been truly invaluable.

I also thank the staff at Nolo Press, including Stephanie Harolde and Steve Elias for their suggestions, and Terri Hearsh for her help in incorporating the drawings.

Table of Contents

4 "Drawing" With a Camera

5 Patent Drawings in General

6 Utility Patent Drawings

7 Design Patent Drawings

8 General Standards

9 Responding to Office Actions

Appendix
Tear-Out Forms

Petition for Submitting Black-and-White Photographs

Petition for Submitting Color Photographs or Drawings

Request for Entry of Drawing Amendment

Index

Introduction

This is a companion book to *Patent It Yourself*, a Nolo Press book that provides step-by-step guidance to you, the inventor, for obtaining a U.S. patent without using a patent attorney or patent agent.

A patent application requires three distinct components:

- a set of detailed drawings showing the invention (see Chapter 5, Section A1, for exceptions to this requirement),
- a written portion that includes a specification and claims—the technical and legal descriptions of the invention—and
- certain formal paperwork, including application forms and a fee.

Although *Patent It Yourself* provides the basic guidelines for making patent drawings, its primarily focus is on the written portion and the formal paperwork of the application. Many readers of *Patent It Yourself* have asked for a more detailed guide to, and explanation of, patent drawings. This is it.

What This Book Adds

The Patent Drawing Book supplements *Patent It Yourself* by providing step-by-step guidance for preparing formal patent drawings (drawings that meet strict PTO requirements), including:

- An introduction to drawings in general (Chapter 1)
- Drawing with pen and rulers (Chapter 2)
- Drawing with a computer (Chapter 3)
- "Drawing" with a camera (Chapter 4)
- An introduction to patent drawings (Chapter 5)
- Different types of patent drawings (Chapters 6 and 7)
- PTO drawing standards (Chapter 8), and
- Responding to PTO examinations or Office actions (Chapter 9).

What This Book Can Do For You

Professional patent draftspersons typically charge $75 to $150 per sheet of patent drawings (each sheet may contain several figures or separate drawings). Most patent applications typically have between 2 to 10 sheets of drawings. By reading *The Patent Drawing Book* and making your own patent drawings, such as the ones shown here, you can save between about $150 and $1,500 per patent application. Once you learn the skills, you can do all the drawings yourself for any subsequent patent applications you file. Furthermore, you will be able to make drawings for a promotional brochure for marketing your invention to prospective manufacturers or customers. Therefore, you may be able to save hundreds, or even thousands, of dollars in the years to come.

You may also be able make the drawings more accurately than a hired professional, because you know your invention best. By doing your own drawings, you do not have to take the time to make someone else understand your invention, or have to send the drawings back and forth for corrections. Also, you will have the great satisfaction of properly completing the entire patent application by yourself—an impressive accomplishment for an inventor. ∎

General Introduction to Drawing

This chapter provides the background information you need to understand the more advanced concepts that are presented in later chapters. Basic drawing principles, including the different types of drawing views and foreshortening (a technique for making realistic views) are presented here. We also provide an overview of several drawing methods, to show you that making patent drawings is probably easier than you may have anticipated.

A. Different Drawing Views

Any physical object can be seen from a great variety of view angles—for example, head-on, from the side, from the top, and from the back. Of course, a single drawing, also known as a drawing view or a figure, may only show an object from one view angle. Typically, a single figure cannot show all of the important features or parts of an object, because some of them may be on an opposite side that is not visible in the view. Therefore, when you need to clearly explain the structure of an invention in a patent application, several drawing views may be necessary to show the object from different angles.

Certain view angles have conventional names, so that they can be immediately understood when referred to. Let's look at the most common of these views.

1. Orthogonal Views

An orthogonal view is one in which the viewer's eyes are centered over a particular side of the object. Put another way, the viewer's line-of-sight is perpendicular, or orthogonal, to such side. A special object—especially created to look different from every side—is shown in Illustration 1.1 to show the possible orthogonal views, which include the following:

Front Side or Front Elevational View: Shows the front side from a viewpoint centered over the front side.

Rear Side or Rear Elevational View: Shows the rear side from a viewpoint centered over the rear side.

Left Side View or Left Elevational View: Shows the left side from a viewpoint centered over the left side.

Right Side View or Right Elevational View: Shows the right side from a viewpoint centered over the right side.

Top Side View or Plan View: Shows the top side from a viewpoint centered over the top side.

Bottom Side View: Shows the bottom side from a viewpoint centered over the bottom side.

Orthogonal views are relatively difficult to understand because they do not convey a sense of depth, so that the shape of many surfaces appears ambiguous. Despite such a shortcoming, orthogonal views are commonly used in patent drawings because they are relatively simple to make. If any of the orthogonal views are considered alone, without the benefit of the other views, the true shape of the object cannot be deciphered. Such ambiguity is shown in Illustration 1.2. An object that appears as a rectangle in an orthogonal view may have many possible true shapes. Therefore, if an orthogonal view does not convey the shape of an object clearly enough, it should also be shown in one or more perspective views.

2. Perspective Views

A perspective view is one that is not orthogonal to or centered over any side. When the view angle is properly selected, it presents a good overall showing of an object as it would be seen in real life by a casual observer. It conveys a good sense of depth, so that it is much easier to understand than orthogonal views. The special object of Illustration 1.1 is shown in typical perspective views in Illustration 1.3, which include the following:

Front Perspective View: Shows the front side somewhat angled away.

Rear Perspective View: Shows the rear side somewhat angled away.

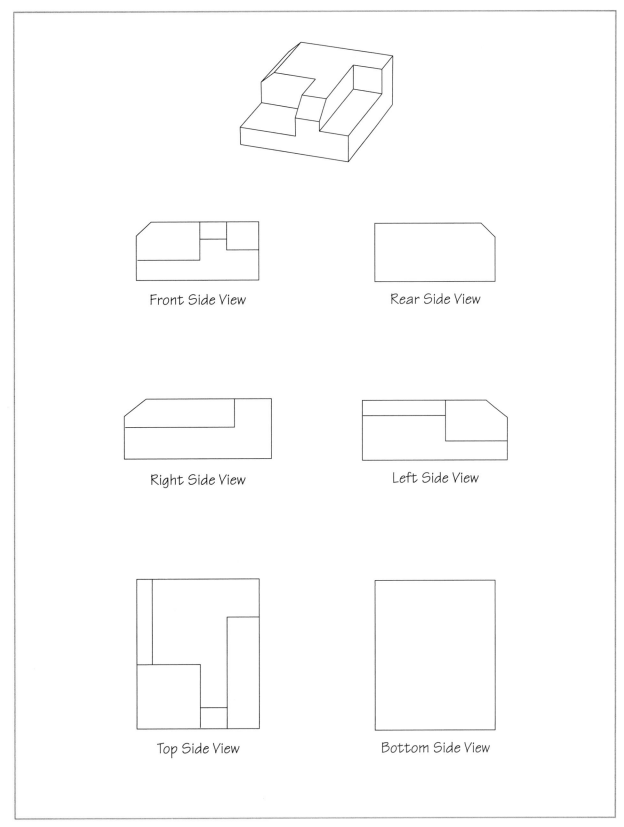

Illustration 1.1—Orthogonal Views

Shape of Orthogonal View

True Shapes

All these shapes appear to be identical in a top view.

Top View

Illustration 1.2—Orthogonal View May Be Ambiguous

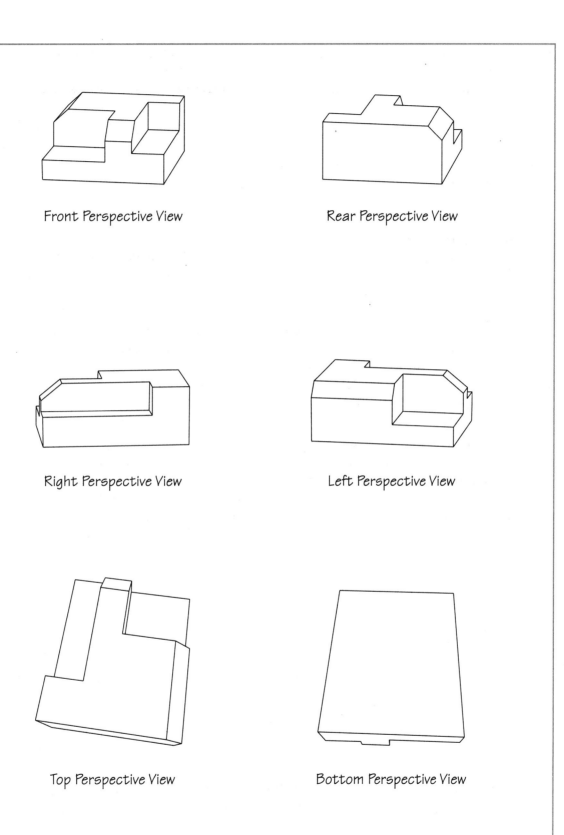

Illustration 1.3—Perspective Views

Right Perspective View: Shows the right side somewhat angled away.

Left Perspective View: Shows the left side somewhat angled away.

Top Perspective View: Shows the top side somewhat angled away.

Bottom Perspective View: Shows the bottom side somewhat angled away.

3. Variations of a Perspective View

If the two sides of an object are equally visible—for example, the top and front—then it may be called either a top perspective or front perspective view. The view angle of any particular perspective view may be varied. Using the same special object of Illustration 1.1, some variations on the front perspective views are shown in Illustration 1.4, which include the following:

Front Perspective View (from above): Shows the front from a higher viewpoint off to one side.

Front Perspective (from the same level): Shows the front from a viewpoint off to one side, but at the same level. Such a view is almost as

ambiguous as the front orthogonal view, so it is not recommended.

Front Perspective (from below): Shows the front from a lower viewpoint off to one side.

One particular type of perspective view is the isometric (iso = equal; metric = measurement) view, from which the viewer's eyes or viewpoint is positioned exactly between three orthogonal views, as illustrated by the simple cube in Illustration 1.5.

Other types of drawing views include the following:

Exploded View: The parts of a device are shown dissembled and spread apart in space to show otherwise hidden features, as shown in Illustration 1.6, a water pipe fitting. Exploded views may be orthogonal or perspective. For example, there can be a front exploded view, a side exploded view, etc.

Sectional View: Part of an object is sliced away to show interior structures. Sectional views may also be orthogonal or perspective. For Illustration, there can be a front sectional view, a side sectional view, a front perspective sectional view, etc. The view shown in Illustration 1.7 is a side perspective sectional view of the water pipe fitting of Illustration 1.6.

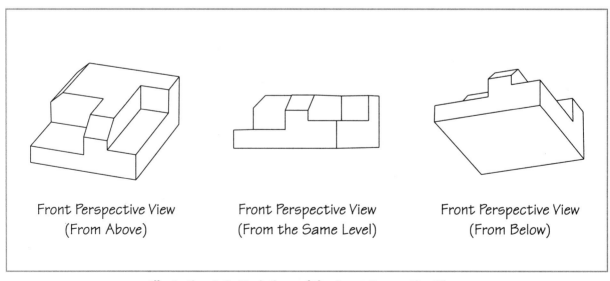

Front Perspective View (From Above)   Front Perspective View (From the Same Level)   Front Perspective View (From Below)

Illustration 1.4—Variations of the Same Perspective View

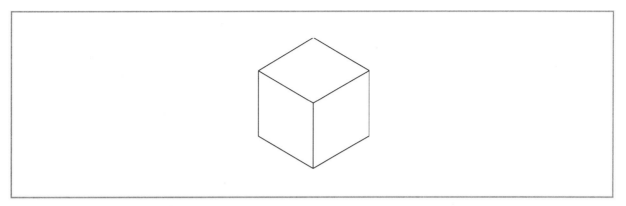

Illustration 1.5—Isometric View

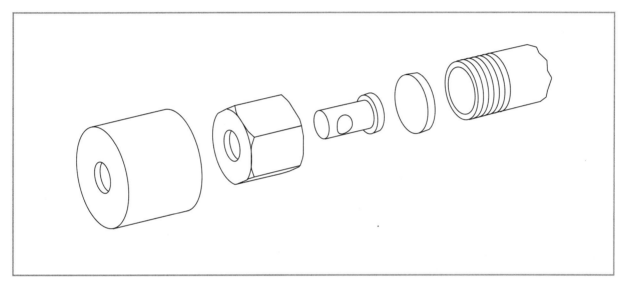

Illustration 1.6—Exploded View

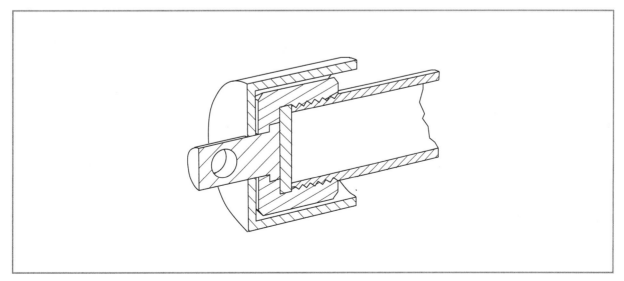

Illustration 1.7—Sectional View

B. Perspective Foreshortening

In real life, objects in the distance appear smaller than similar objects up close. The same principle also applies to a single, three-dimensional object: its far end appears smaller than its near end, and its parallel edges appear to converge. The closer you are to the object, the greater the effect appears. To see the effect very clearly, put a long rectangular object, such as a toothpaste box, very close to your eyes. You will notice that its far end appears much smaller than its near end, and that its parallel edges appear to converge.

The technique of representing such an effect in a drawing is known as foreshortening. It is applied to perspective views to make them more realistic. The next time you go to a museum or look at an art book, compare medieval paintings, which were done without foreshortening, to Renaissance paintings, which were done with foreshortening. You will see that the medieval paintings appear flat and somewhat cartoon-like, whereas Renaissance paintings are much more realistic representations of people and things.

The degree of foreshortening is inversely proportional to the viewing distance. That is, an object seen from a short distance is drawn with more foreshortening, and an object seen from a greater distance is drawn with less foreshortening. Using the same toothpaste box, you can see that its appears highly foreshortened when it is very close to your eyes, and not foreshortened at all when it is far away.

1. No Foreshortening vs. Excessive Foreshortening

Illustration 1.8 shows a square box drawn without foreshortening, and also with different degrees of foreshortening. The box drawn without foreshortening represents its appearance as seen from a great distance. Its parallel edges are drawn as perfectly parallel lines, so this view is also known as a parallel view. Without foreshortening, the box actually appears slightly distorted.

The box drawn with excessive foreshortening represents its appearance as seen from an extremely short viewing distance, such as when it is positioned right up against your eyes. Although excessive foreshortening causes the box to appear greatly distorted, it clearly shows how foreshortening is applied: the parallel edges of the box are drawn as converging lines that, if extended, will intersect at points known as "vanishing points." The vanishing points on the sides lie on the horizon (a horizontal line), and the central vanishing point lies below "ground level." In the illustration shown, the box is being seen from above.

2. Realistic Foreshortening

The box is most realistically illustrated with normal foreshortening, that is, a small degree of foreshortening, which represents the object as seen from a normal viewing distance. The parallel edges are drawn as slightly converging lines. The vanishing points of such slightly converging lines are far off the page, so they are not shown.

The rule of thumb is that the greater the foreshortening, the closer the vanishing points are positioned, and the lesser the foreshortening, the farther the vanishing points are positioned. A drawing without foreshortening has no vanishing points, because parallel edges are drawn as parallel lines, which do not converge.

Drawings done with normal foreshortening are the most realistic, but drawings done without foreshortening (parallel views) are perfectly acceptable for patent drawings. Foreshortening is a difficult technique to apply with pen and rulers, but as discussed in Chapter 3, Section D3, it is extremely easy to apply with a computer. You can ignore foreshortening in patent drawings, but you can make much more attractive marketing brochures if you use it.

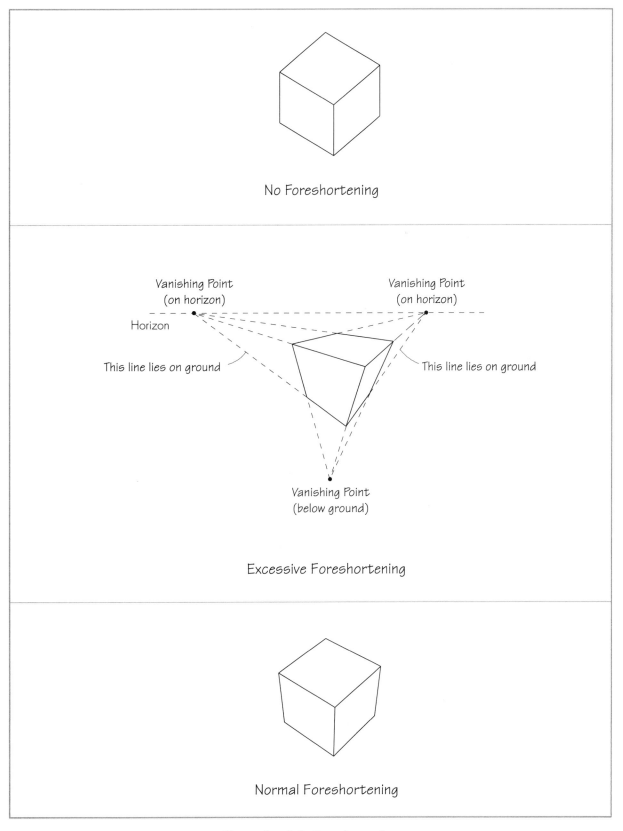

No Foreshortening

Vanishing Point
(on horizon)

Vanishing Point
(on horizon)

Horizon

This line lies on ground

This line lies on ground

Vanishing Point
(below ground)

Excessive Foreshortening

Normal Foreshortening

Illustration 1.8—Foreshortening

C. Making Drawings With Pen and Rulers

There are several methods for making patent drawings. The traditional or old way of making patent drawings is with pen and rulers. A set of basic tools for this method can be assembled relatively inexpensively, and making simple drawings is fairly easy. However, with pen and rulers, there is little room for mistakes, because except for very small marks, it is very difficult to correct misplaced ink lines. Nevertheless, with careful planning of drawing positioning (layout), and great care in laying down ink lines, drawing with pen and rulers is still a viable technique. In fact, according to one survey, most professional patent draftsmen still make drawings this way.

1. Necessary Tools

The necessary tools include pencils for preliminary sketches, ink drafting pens for drawing ink lines, rulers for making straight lines, triangles for making angled lines, templates for making certain standard shapes, curve rules for drawing curves, an optional drafting table, and high-quality ink drawing paper.

2. Pen and Rulers Drawing Techniques

Pen and rulers may be used to make patent drawings in the following ways:

Drawing From Scratch: You can draw an object by visualizing in detail what it should look like, carefully sketching that image on paper with a pencil, correcting it until it looks about right, and finally inking in the lines. Drawing from scratch requires some basic drawing skills. If you need to learn such skills, you can do some additional reading as suggested at the beginning of Chapter 2, or you can draw with a computer instead, which eliminates the need for traditional drawing skills.

Tracing: Tracing is much easier than drawing from scratch. An obvious method is to trace a photograph of an object that you wish to draw, as

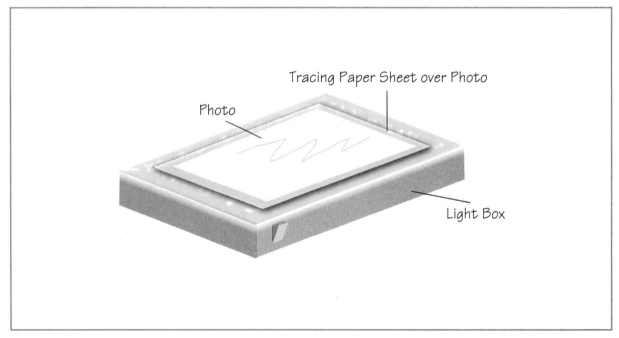

Illustration 1.9—Tracing a Photo

shown in Illustration 1.9. An actual, three-dimensional object can also be traced by using a device called Camera Lucida, available at art supply stores, which projects an image of the object onto a drawing surface. As shown in Illustration 1.10, you may also trace an actual object—again, we use a simple box to illustrate—by positioning a transparent drawing sheet on a transparent sheet of glass or acrylic, looking at the object through the glass, tracing the lines of the object on the drawing sheet, and photocopying the tracing onto a sheet of paper. This technique is discussed in greater detail in Chapter 2, Section C3. Tracing requires very little skill other than a steady hand.

Drawing to Scale: You can also draw by scaling—that is, reducing or enlarging—the dimensions of an object to fit on a sheet of paper, and draw the lines with exact scale dimensions. For example, if an object has a height of 20 inches and a width of 12 inches, you can reduce those dimensions by 50%, so that you would draw it with a height of 10 inches and a width of 6 inches on paper, as shown in Illustration 1.11. All other dimensions of the object are scaled accordingly for the drawing. Making a drawing that looks right is easier by drawing to scale than by drawing based on only a mental image.

The techniques and tools for drawing with pen and rulers are explained in greater detail in Chapter 2.

D. Drawing With a Computer

CAD (computer-aided drafting or design) has been around since perhaps the 1970s. Initially, CAD was a tool found only in companies that could afford the expensive mainframe or workstation computers. With the introduction of personal computers in the early 1980s, CAD became a much more affordable tool. However, early CAD software for personal computers was difficult to use. Also, it ran slowly because of the anemic processing power of early personal computers. In the 1990s, the rapidly-increasing speed, affordability, and ease of use of personal computers and software have enabled vast improvements in productivity across many different fields, including CAD. By the mid 1990s,

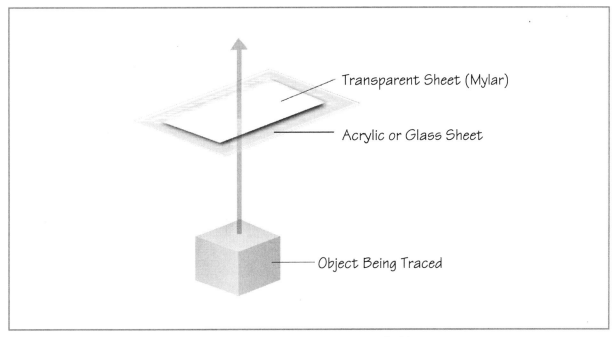

Transparent Sheet (Mylar)

Acrylic or Glass Sheet

Object Being Traced

Illustration 1.10—Tracing an Actual Object

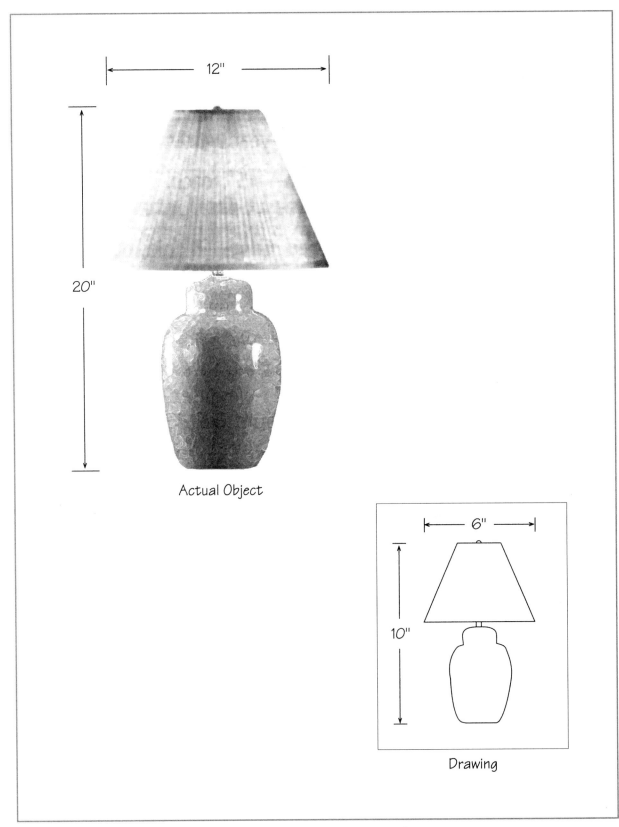

12"

20"

Actual Object

6"

10"

Drawing

Illustration 1.11—Drawing to Scale

personal computers have become a common appliance in many homes. Also, CAD software, which ranges from dirt cheap to very expensive, has become much more powerful and easy-to-use.

CAD allows you to produce accurate drawings even if you consider yourself to have little or no artistic ability. In fact, no drawing skills in the traditional sense are needed at all. Furthermore, CAD enables you to correct mistakes as easily as a word processor enables you to edit words in a document. Even if you discover a mistake after you print a drawing, you can easily correct the mistake and print a new copy. To use CAD, you will need some computer skills, but if you know how to type letters on your computer, you can easily learn how to draw with it.

1. Equipment

You will need either a PC (IBM compatible) or a Mac, an ink jet or laser printer, a CAD program, an optional scanner, and an optional digital camera.

2. Computer Drawing Techniques

A computer may be used to make patent drawings in the following ways:

Tracing: If you have a scanner, you can scan a photograph of an object, import (load) the scanned image into a CAD program, and trace it easily, as shown in Illustration 1.12, a photo of an aircraft (the black outlines are the tracing lines—difficult to

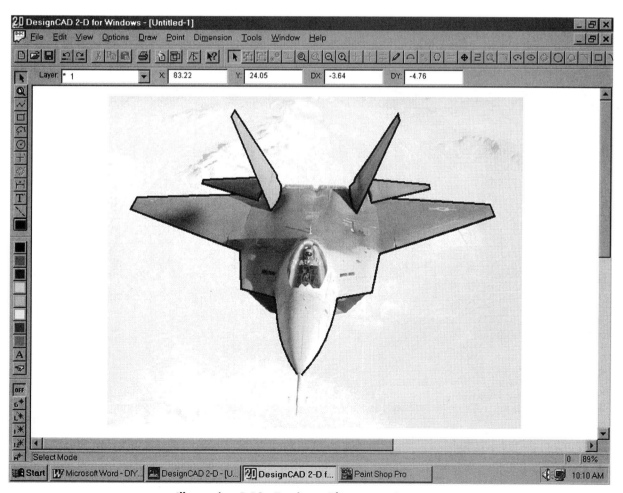

Illustration 1.12—Tracing a Photo on a Computer

see in a black-and-white book). If you have a digital camera you can take a photograph of the object and download (transfer) the image directly into your computer through a cable, without having to print and scan the photograph. Once it is in your computer, tracing the image is very easy. Since you use a mouse instead of an ink pen, you don't even need a steady hand.

Drawing From Scratch: A CAD program will enable you to construct an accurate, three-dimensional (3D) representation model of your invention within the computer, such as the pipe fitting shown in Illustration 1.13. A 3D model is typically built by using and modifying basic geometric building blocks, such as boxes, cylinders, planes, and custom-defined shapes. You may create each part with specific dimensions, or you may simply draw a shape that looks about right. You can easily rotate the finished model to see it from any angle. You can also easily zoom in or out to adjust the viewing distance, which is equivalent to adjusting the degree of foreshortening (in CAD, the term "viewing distance" is used instead of "foreshortening"); see Chapter 3, Section D3. Once you are satisfied with the view, you can print it as a line drawing (a drawing comprised of dark lines

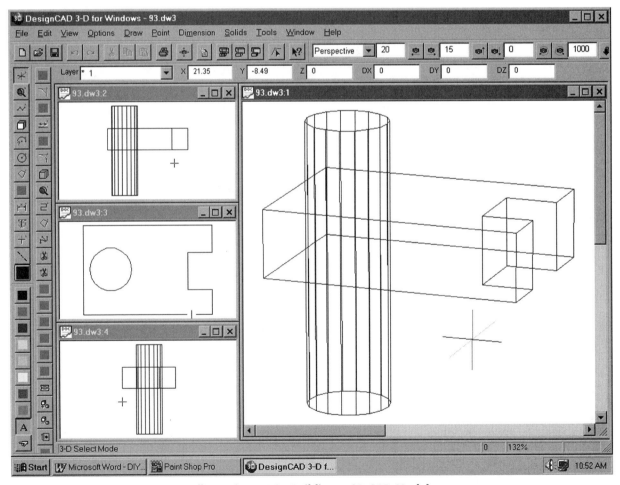

Illustration 1.13—Building a 3D CAD Model

on a light background). Therefore, you can make wonderful looking drawings with a computer, even if you consider yourself to be a terrible artist.

These techniques and tools are explained in greater detail in Chapter 3.

E. "Drawing" With a Camera

Almost everyone has some degree of familiarity with photography. Obviously, a camera can take an accurate photograph or "drawing" of an object. Photographs may also be submitted as patent drawings, subject to certain requirements that will be explained in Chapter 4, or they can be converted into line drawings by tracing them. Although photography spares you from having any drawing skills, you must have some photographic skills to take clear pictures, including a basic understanding of lighting and exposure.

1. Equipment

To take an accurate photograph, you will need a 35 mm camera with selectable aperture, zoom and macro (close up). You will also need a tripod.

2. Taking Pictures

There are mountains of books on photographic techniques, and cameras typically come with booklets on basic techniques, so we will not go into great details. We will cover a few simple techniques in Chapter 4 that will enable you to take pictures good enough for filing as patent drawings.

3. Tracing Pictures

A photograph can be converted into a line drawing by putting a piece of paper over it and tracing it with a pencil, or by scanning it into a computer and tracing it with a CAD program.

F. Summary

As you can see, there are several ways to make patent drawings. If you favor one of them, you may go directly to the chapter that discusses it in detail. Otherwise, a reading of the following chapters will help you select a technique that is right for you. ■

Drawing With Pen and Rulers

This chapter provides general instructions for making drawings with pen and rulers. However, it is beyond the scope of this book to go into great detail on basic drawing skills, which are already covered in many other books.

Additional Reading

For basic drawing skills, we suggest:
- *Basic Drawing Techniques*, written by Greg Albert and Rachel Wolf (North Light Books).
- *Keys to Drawing*, written by Bert Dodson (North Light Books).
- *Learn Cartooning and Drawing the Easy Way*, written by Bruce Blitz (Art Products, Inc.).

These books are available at Flax Art & Design (see Section A, below).

The advantage of making drawings with pen and rulers is that the tools are relatively inexpensive. The drawback is that, because ink marks cannot be easily corrected, there is little room for making mistakes. If you can be very careful in laying down ink lines, drawing with pen and rulers is still a viable technique.

A. Necessary Tools and Supplies

All the traditional drawing tools and supplies you will need can be found at art supply stores. If there is no such store in your area, you can request catalogs from Flax Art & Design (415-552-2355) and California Art Supply (510-832-3121), which are large art supply stores in San Francisco and Oakland, California. They carry slightly different lines of products, so get both catalogs to compare selection and prices.

Generally, drawings of physical objects require the greatest number of tools because of their complex lines, whereas drawings of graphical symbols, such as flowcharts and electrical schematics, require fewer tools because of their simple, regular shapes. You will not need a complete set of drawing tools; what you need depends on what you want to draw and how you want to draw it. You can determine the tools and supplies you need after you have read this chapter. You also may try drawing in pencil after reading this chapter to see if you can produce satisfactory drawings before you invest in new tools and supplies.

1. Essential Tools and Supplies

The following tools and supplies are essential for traditional, manual drawing techniques.

1. **Pencil, either wood or mechanical, for making light, erasable sketches.** A soft lead pencil, such as 2B, is usable as long you draw lightly with it. A medium lead pencil, such as HB or F, is probably more suitable. A hard lead pencil, such as H or above, is generally not suitable, because too much pressure is needed to make a visible line, and the pressure will tend to score grooves in the paper.

2. **Soft or kneaded eraser for erasing pencil marks without harming paper.** A good brand of soft eraser is Staedtler, by Mars Plastic. Soft erasers are easier to handle, but they leave debris on the drawing that must be brushed away. Kneaded erasers, which are usually only available at art supply stores, are pliable and do not leave debris. They must be kneaded, like a dough, to push in the dirtied parts and bring the clean inner parts to the surface. You may also wish to get an electric eraser, which is handy for erasing large areas.

3. **Technical pens for drawing ink lines.** Technical pens are ink pens made specifically for precision ink drawings. They come in

different sizes; you should have at least the 0.13 mm and 0.25 mm sizes. Well known brands include Koh-I-Noor Rapidograph and Rotring Rapidograph, which cost about $20 each. Alternatively, good quality extra fine to medium point felt-tip or plastic-tip pens, which are only about $2-$3 each, may be used. A good brand is Tech-Liner Pens by Alvin. Do not use ball-point pens, roller-ball pens, or fountain pens, because they do not make lines that are sharp and black enough to meet Patent and Trademark Office standards.

4. **Non-clogging black ink for the pens, unless you use the pre-filled models.** Some good brands include Higgins Black Magic and Rapidograph Ultradraw.

5. **Bottle of white correction fluid with a pen-type applicator.** A good brand is Pentel Quick Dry Correction Fluid. Avoid those with brush applicators, because the fluid in them tends to dry out and thicken over time, and becomes difficult to apply.

6. **Drafting board with a built-in parallel rule and protractor (a ruler that slides in a parallel fashion up-and-down or left-and-right on the drafting board for accurately drawing parallel lines, vertical lines, horizontal lines, or lines at any other angle).** This is an essential tool. Rotring sells an inexpensive model (about $60) under the trademark Koh-I-Noor Portable Drafting System. Alternatively, a T-square (a T-shaped device for sliding up and down or left and right along the edges of a drafting board) and a plain, rectangular drafting board may be used. You may also dispense with the drafting board by using the corner of a table with a smooth surface to guide the T-square. If you use a T-square, you will also need a set of triangles and a protractor for drawing lines of different angles. The overall cost of assembling these components may not be lower than the Koh-I-Noor Portable Drafting System, depending on the quality of the components, and they are definitely more

difficult to use, because you have to keep the T-square and the triangle aligned properly while you control the pen at the same time.

7. **1/8" and 1/4" lettering guides (templates) for writing text.** Alternatively, transfer type (rub-on lettering) of a simple typeface or style in the same sizes may also be used. The guides cost about $5 each, and transfer type costs about $13 per sheet.

8. **Post-It® brand removable tape or masking tape for taping down drawing paper on the drafting board.** Only use tape that does not damage paper.

9. **Parchment tracing paper for tracing photographs or sketches.** Any brand will do.

10. **Vellum for finished drawings.** Vellum is a tough, matte (frosted), translucent paper that takes ink very well, and can be repeatedly erased without damage. A proper ink drawing paper must be used, because other papers will cause the ink to feather—that is, seep between the paper fibers—and spread out. Vellum costs about $18 for 100 sheets.

2. Additional Tools Recommended for Drawing Physical Objects

In addition to the tools listed above, there are some additional tools that are useful for drawing physical objects:

1. **Engineer's triangular scale (a six-sided ruler) for drawing to scale.** One side of the triangular scale is marked with full-size inches with ten divisions each, and other sides are marked with 1/2, 1/3, 1/4, 1/5, and 1/6 scale inches. For example, on the 1/2 scale side, each inch mark is 1/2 the size of an actual inch. They are as low as $5 each.

2. **Various circle templates (plastic sheets with holes) for drawing circles of different fixed sizes.** About $7 each.

3. **Various ellipse templates for drawing ellipses of different fixed sizes and**

shapes. An ellipse is the shape of a circle when seen at an angle. About $7 each.

4. **30 and 45 degree triangles with beveled edges.** The beveled edges are necessary for preventing ink from seeping under them. They are as low as $4 each.

5. **Compass capable of holding a technical pen for drawing circles and arcs of custom sizes.** About $25.

6. **Ellipsograph for drawing ellipses of custom sizes and shapes.** An ellipsograph is a tool for drawing custom-sized ellipses. Although it will enable you to draw an ellipse with the exact shape desired, it is more difficult to use than an ellipse template, because it has to be carefully adjusted and positioned to get the desired results. It has reportedly been discontinued by the manufacturer, but some vendors may still have them in stock.

7. **Various French curves (templates with many curved edges of fixed shapes) for drawing uneven curves.** French curves require a lot of practice to use, because it is often difficult to find an edge with the desired curvature. About $8 for a set.

8. **Flexible (adjustable) curve for drawing custom-shaped curves.** A flexible curve may be used instead of a French curve, but you will have to adjust it every time you need a different shape. About $6.

9. **Transparent grid overlay, which is a transparent plastic sheet with a grid etched on it for plotting difficult shapes.**

10. **Preprinted grid paper.** Available in various sizes, including inch, metric, isometric, and perspective. Used under a sheet of tracing paper to serve as a guide.

3. Additional Tool for Drawing Graphical Symbols

To draw graphical symbols, you may want to have the following tool:

1. **Suitable symbol templates for drawing symbols**. Templates are available for a variety of specialized symbols, such as electronics, architectural, and flowchart symbols.

4. Additional Tools for Tracing Photographs or Sketches

For tracing photographs or sketches, you may want these additional tools:

1. **Light box**—a box with internal light and translucent white top. About $80.

2. **Clear polyester film.**

The pantograph (a mechanical parallelogram device) is also available for tracing drawings. It may be adjusted to make a drawing at various scales to the original. However, it produces very inaccurate results, so it is not recommended. A photocopier may be used to make enlargements or reductions much faster and more accurately.

5. Additional Tools for Tracing Actual Objects

You may want these additional tools for tracing actual objects:

1. **Camera Lucida**—a clever device that includes a lens mounted on an adjustable arm for projecting an image of an actual object onto a drawing surface at selectable magnification or reduction, so that the image size on the paper may be adjusted. Interestingly, this device was invented back in 1807. It is available at Flax Art & Design (see Section A, above) for several hundred dollars. It is relatively difficult to use.

2. **A home-made, direct-view tracing device that includes a 12" x 24" sheet of 1/8" thick transparent acrylic or Plexiglass®, and a support device for supporting the transparent sheet above the object being traced.** The acrylic sheet is available at plastic supply stores, such as Tap Plastics

(for catalog, call 415-829-4889). Avoid using glass, which may break and cause injuries. There is no commercially-available supporting device suitable for this purpose, so you will have to create your own with your inventive ingenuity. Section C, below, provides some suggested solutions.

3. **Clear polyester film** for use with the homemade, direct-view tracing device (not the Camera Lucida, which projects an image onto paper).

6. Total Cost of Tools

The cost of a set of tools and supplies may range from as little as $100 to several hundred dollars, depending on your particular needs, and how frugal you are. The cost is roughly equivalent to the cost of a CAD (computer aided drafting) program. If you already have a computer, you may want to read Chapter 3, which deals with computerized drafting, to see if you prefer to use the computer instead. If you do not have a computer, then drawing with pen and rulers is the most economical way to go.

B. Basic Drawing Rules and Techniques

Everyone planning to make a patent drawing with pen and rulers should review the rules and techniques described in this section.

1. A Checklist of Rules and Techniques

To produce good drawings, you will need to apply the following basic rules and techniques:

- Always sketch a drawing lightly in pencil first, then apply ink lines over the final pencil marks.

- All ink lines should be drawn with the aid of guides, such as rulers, templates, and French curves. Use freehand drawing only when there is no alternative.

- Position the pen substantially vertically, and apply even pressure when moving it for a smooth and even line. Do not tilt the pen when you move it; otherwise the line width will be uneven.

- Avoid going over a line for a second pass; otherwise the line will become too thick or uneven due to the slightly changed position of the pen on the second pass.

- Position the pen so that its very tip does not touch the edge of rulers and other guides; otherwise the ink will touch and spread under the guides.

- After drawing a line, pull or lift the guide from the line without crossing over it; otherwise it will smear the ink.

- Wait patiently until ink lines dry completely before putting anything over them, erasing pencil marks near them, or erasing misplaced ink lines; otherwise you will smear the ink. Generally, ink takes a few seconds to a minute to dry completely. Look closely at the ink lines; if they are shiny, they are still wet.

- Avoid erasing ink lines as much as possible, because erasing roughens the paper. Subsequent ink lines drawn over the roughened area will feather or bleed out.

- Use white correction fluid for covering unwanted ink marks. Make sure the ink is dry before applying the correction fluid. Dab the fluid on quickly and sufficiently so you don't have to go over the same area again, because that tends to roughen the fluid as it dries, so that it will not take ink well and may show up as blotches in photocopies. Do not rub the fluid on, because the tip may mix the ink with the fluid and darken the fluid.

- Lines that meet to form sharp corners must touch precisely without overlap, as shown in Illustration 2.1.

- Rounded corners should be drawn by drawing the curved segment first, then drawing the straight lines from the ends of the curved segment, as shown in Illustration 2.2.

- Erase pencil marks by holding down the paper with one hand, and stroking an eraser along the paper away from the hand holding the paper. Do not stroke the eraser back and forth, because—unless the paper is secured all around—the eraser will push the paper toward the hand holding it and wrinkle it.

- Make full use of the space on each sheet of paper to make a drawing as large as necessary to show all details clearly.

- Unless you are an expert calligrapher, always use a lettering set to write text, or use transfer type for applying lettering to drawings. If you use a lettering set, use a pencil line for aligning the letters, as shown in Illustration 2.3. Do not use a pencil line for aligning transfer type, because the line cannot be erased later without also rubbing off the letters. Misplaced transfer type may be lifted off with tape, but be careful not to damage the paper. Be careful not to rub off the letters after they are applied.

- Do everything with great care and patience.

2. Practice, Practice, Practice

If you are unfamiliar with doing high-quality ink drawings, you should not make your drawing project your first ink drawing experience. If you do, you will make mistakes right from the start, get discouraged, and give up. Instead, you should practice drawing basic shapes—such as straight lines, rectangles, circles, and corners—with the techniques presented above, so that you become familiar with the tools and medium (ink and paper). Reading one of the books recommended at the beginning of the chapter will be helpful. Next, you should practice drawing more complex shapes, such as the more difficult portions of the drawing

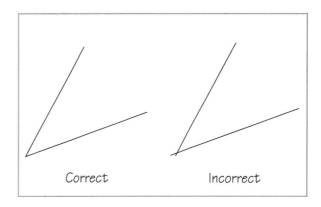

Illustration 2.1—Lines Forming Corners

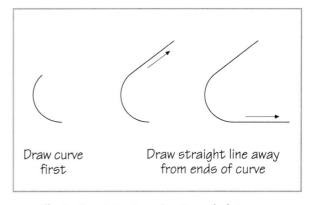

Illustration 2.2—Drawing Rounded Corners

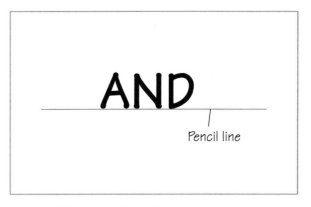

Illustration 2.3—Applying Lettering

you ultimately want to do. By becoming familiar with the tools and techniques first, you will avoid becoming discouraged before you start your project.

C. Tracing Photographs and Objects

You can produce accurate drawings with relative ease by either tracing photographs of an object, or by tracing the actual object.

1. Tracing Photographs

To trace a photograph or other printed graphics, follow these steps:

1. Take close-ups of the object. It should appear as large as possible in the photos. Make 4" x 6" or larger prints, so that the image of the object is large enough for producing a suitably-sized drawing. See Chapter 4 for details on how to take suitable photographs.

2. Tape the photograph on a light box, and tape a piece of vellum over it, as shown in Illustration 2.4.

3. Trace the photograph very carefully and lightly with a pencil. Avoid making dark lines or pressing too hard and scribing grooves in the paper.

4. Mount the paper on a drafting board, ink over the pencil lines, and erase the pencil marks. Always draw ink lines with the aid of rulers and templates. When inking circles, ovals, or curves, select a template that most closely matches the shape desired.

5. After the ink is completely dry, erase any pencil marks that remain visible outside the ink lines.

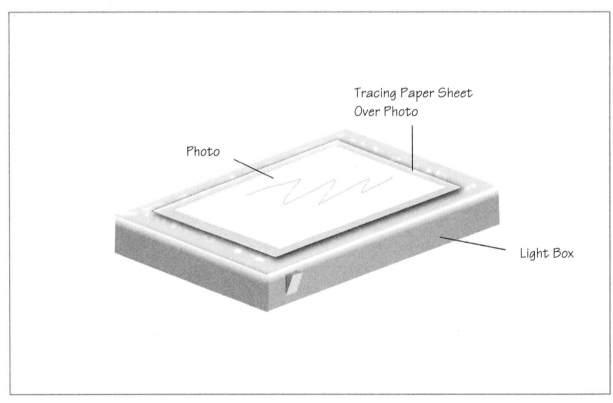

Photo

Tracing Paper Sheet Over Photo

Light Box

Illustration 2.4—Tracing a Photo

2. Tracing an Object With a Camera Lucida

To trace an actual object with a Camera Lucida, follow these steps:

1. Mount the arm on a table, and position the lens over a sheet of paper, as shown in Illustration 2.5.

2. Position the object on the table.

3. Look through the lens at the paper with one eye, and adjust the positioning of the lens and the object until an image of the object appears on the paper.

4. Adjust the distance of the object, or change lens to adjust the image size on the paper.

5. Trace the image lightly with a pencil. You may trace freehand, or with the aid of rulers and templates. Keep your head steady to keep the image positioned on the paper in precisely the same position at all times. Every time you move your head away from the lens and go back later, you must realign the image with the lines on the paper.

6. After you finish tracing, mount the paper on a drafting board, and ink over the pencil lines. Always draw ink lines with the aid of rulers and templates. When inking circles, ovals, or curves, select a template that most closely matches the shape desired, or use an adjustable curve. If you traced freehand, the pencil lines would be somewhat crooked, so you should apply the ink lines for a best fit.

7. After the ink is completely dry, erase any stray pencil marks that remain visible outside the ink lines.

Note: *Tracing with the Camera Lucida is relatively difficult, because you must keep your head very still while your hand moves. Tracing a photograph is much easier.*

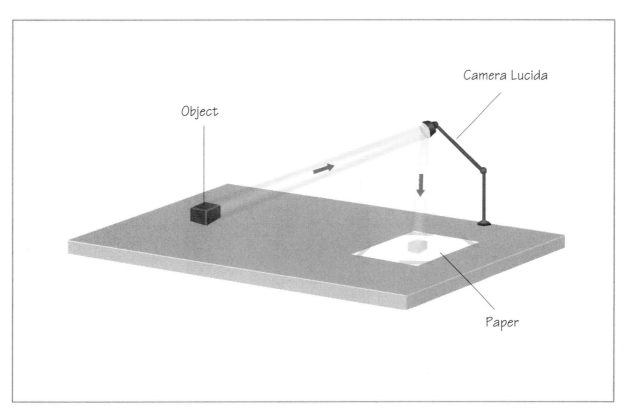

Illustration 2.5—Using a Camera Lucida

3. Tracing an Actual Object With a Direct-View Device

To trace an actual object with a direct-view device, as discussed in Section A5, above, follow these steps:

1. Position the object on the ground, and orient it for a suitable view (front, side, top, etc.) when seen from above. Use tape, blocks, modeling clay, earthquake wax, or whatever is suitable to fix it in the desired position, as shown in Illustration 2.6.

2. Tape a sheet of polyester film on a sheet of transparent acrylic.

3. Position the acrylic sheet over the object and adjust the height or distance of the sheet so that the object fits within the desired drawing area of the polyester film. There is no commercially-available device for supporting the acrylic sheet for this purpose, so you will have to use your ingenuity to devise a solution. One possibility is to support the acrylic sheet between two tables of identical height, and position the model below the acrylic sheet. The object can be lifted off the ground with a stack of magazines or encyclopedias to adjust its distance from the acrylic sheet. Another possibility is to support the sheet

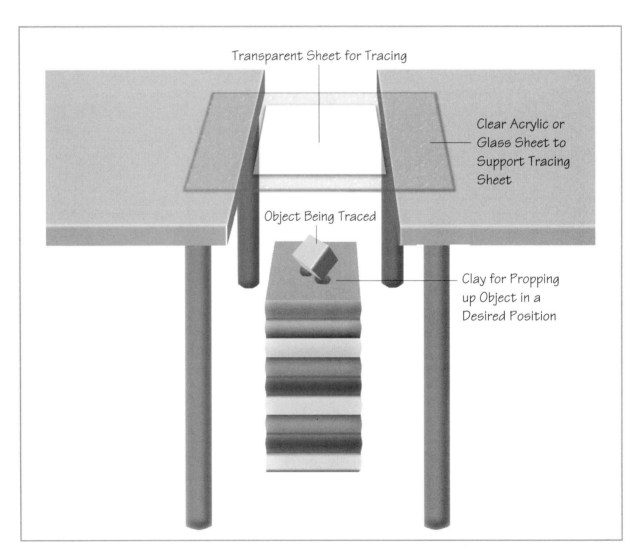

Transparent Sheet for Tracing

Clear Acrylic or Glass Sheet to Support Tracing Sheet

Object Being Traced

Clay for Propping up Object in a Desired Position

Illustration 2.6—Tracing an Actual Object

between the heads of two camera tripods, which can be easily adjusted to support the sheet at a range of different heights. For a larger object that requires a greater distance to fit within the tracing film, the acrylic sheet may be positioned vertically and spaced horizontally from the model. This may be done on a long table, as shown in Illustration 2.7, or with tripods, as shown in Illustration 2.8.

4. Close one eye. Position yourself so that you see the object positioned within a desired part of the polyester film. Select two opposite corners of the object, and mark them. Whenever you are tracing, you must keep the same eye closed and see with the other eye, otherwise parallax will cause substantial errors in the drawing.

5. Trace the model very carefully with a pencil. You must keep your head steady to keep the object lined up with the marks whenever you are tracing. Whenever you move away from and later come back to the model, you must adjust your head's distance and lateral position to line up the object with the marks again.

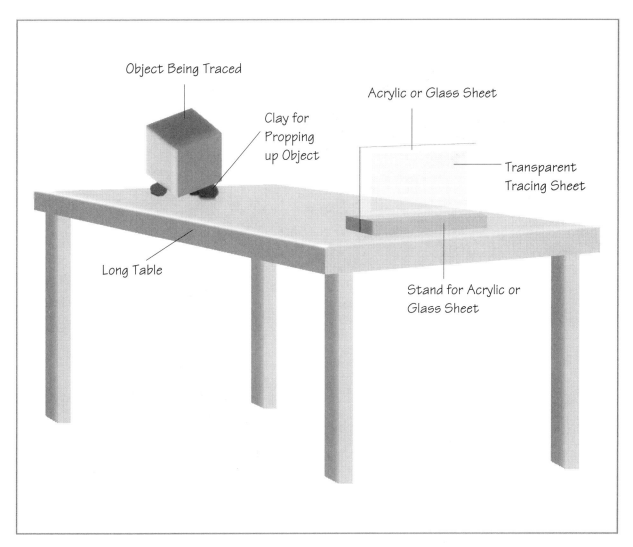

Object Being Traced

Clay for Propping up Object

Acrylic or Glass Sheet

Transparent Tracing Sheet

Long Table

Stand for Acrylic or Glass Sheet

Illustration 2.7—Tracing a Large Object on a Long Table

6. After tracing is completed on the polyester film, tape it on a light box, tape a sheet of vellum over it, as shown in Illustration 2.4, and lightly trace the drawing onto the vellum with a pencil. Be careful not to press too hard and scribe grooves into the paper.

7. Mount the vellum on a drafting board, and ink over the pencil lines. Always draw ink lines with the aid of rulers or templates. When inking circles, ovals, or curves, select a template that most closely matches the shape desired. If you traced freehand, the pencil lines would be somewhat crooked, so you should apply the ink lines for a best fit.

8. After the ink is completely dry, erase any pencil marks that remain visible outside the ink lines.

Note: *Although it is a very inexpensive technique, tracing with the direct-view device is about as difficult as using the Camera Lucida, because you must also keep your head very still while your hand moves. Tracing a photograph is much easier, but relatively more expensive.*

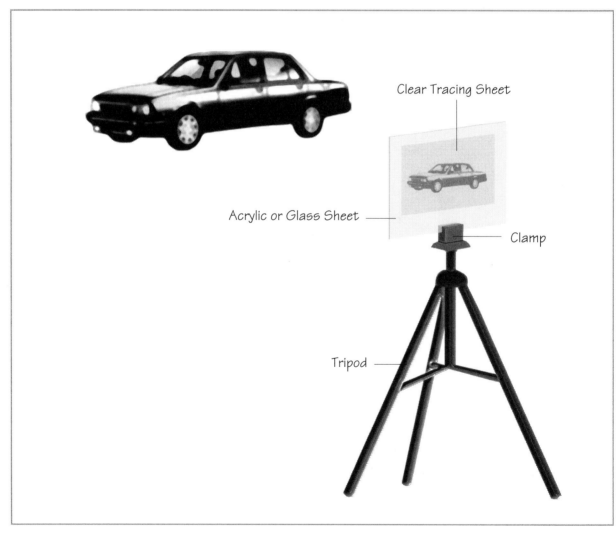

Clear Tracing Sheet

Acrylic or Glass Sheet

Clamp

Tripod

Illustration 2.8—Tracing a Very Large Object

4. Enlarging or Reducing a Tracing

In patent drawings, each figure should be large enough to show essential details clearly without crowding, but should not be unnecessarily large. If your tracing is not of a suitable size, you may enlarge or reduce it with the following method:

1. Enlarge or reduce the tracing with a photo-copier. You can enlarge an enlargement or reduce a reduction to make even bigger or smaller copies, respectively.
2. Mount the photocopy on a light box, and mount another sheet of paper over it.
3. Trace the photocopy lightly in pencil.
4. Mount the second tracing on a drafting board and ink over the lines. Always draw ink lines with the aid of rulers or templates. When inking circles, ovals, or curves, select a template that most closely matches the shape desired. If you traced freehand, the pencil lines would be somewhat crooked, so you should apply the ink lines for a best fit.
5. After the ink is completely dry, erase any stray pencil marks that remain visible outside the ink lines.

D. Drawing From Your Imagination

If the object you wish to draw does not exist, so that you cannot base the drawing on its actual shape as you can when tracing, you will have to base the drawing solely on a mental image.

1. Sketching the Figures

The first step in creating a drawing from your imagination is sketching the figures:

1. Use a sheet of vellum, and start by lightly sketching a rough shape of the object you have in mind, as shown in Illustrations 2.9 and 2.10. For long objects, such as airplanes, human limbs, table legs, etc., start by drawing centerlines, then add the outline around the centerlines. Don't worry about the details. Concentrate on getting the proportions and perspective right. If you fuss with the details, you will lose sight of the overall proportion and shape of the object and come up with something distorted. The pencil marks must be light enough so that they can be completely erased later.
2. Sketch the major features of the object to refine the drawing a little more.
3. Fine-tune the lines and add in small details.

This technique is akin to seeing an image through a pair of adjustable eye glasses that are initially out of focus for you. At first, you can only see a rough shape, but as you adjust the glasses, the image slowly becomes sharper to reveal more of its shape, until the glasses are perfectly adjusted and you can see all the details clearly.

2. Study Similar Objects for Clues

If you have trouble making a drawing look right, study an object that is shaped similarly to the one you wish to draw, or the portion of the object you are drawing. For example, if you have trouble drawing a box in perspective, find a similar box, position it in the same view angle that you are drawing, and study its lines to get an idea of what they should look like, as shown in Illustration 2.11.

When studying the lines, look for qualities such as their angles relative to the vertical or horizontal, angles relative to other lines on the same object, and relative length of each line with respect to other lines on the same object. If you are careful in making these observations, you should be able to create satisfactory drawings. However, if you are still having a great deal of difficulty, you may consider making a model and tracing it, as described in Section C, above, or using a computer. (See Chapter 3 on drawing with a computer.)

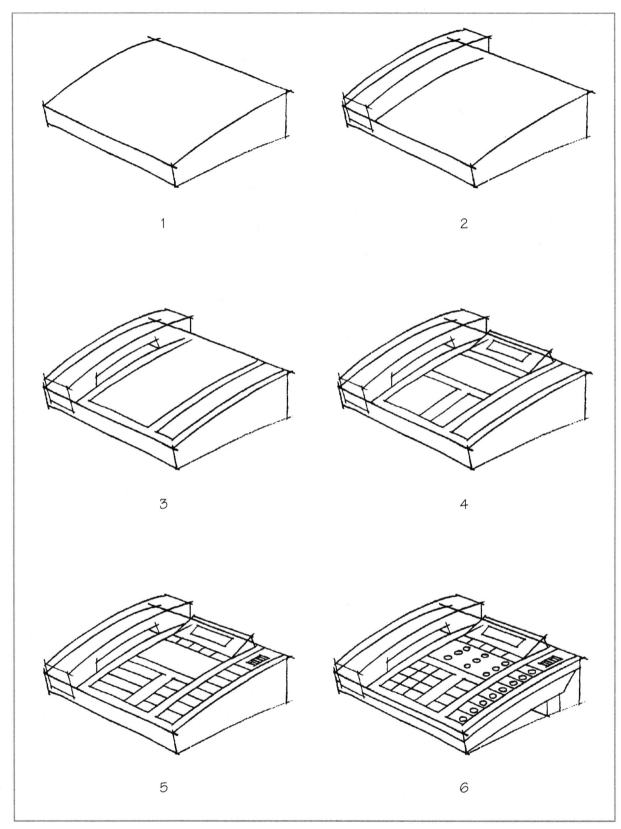

Illustration 2.9—Sketching Technique

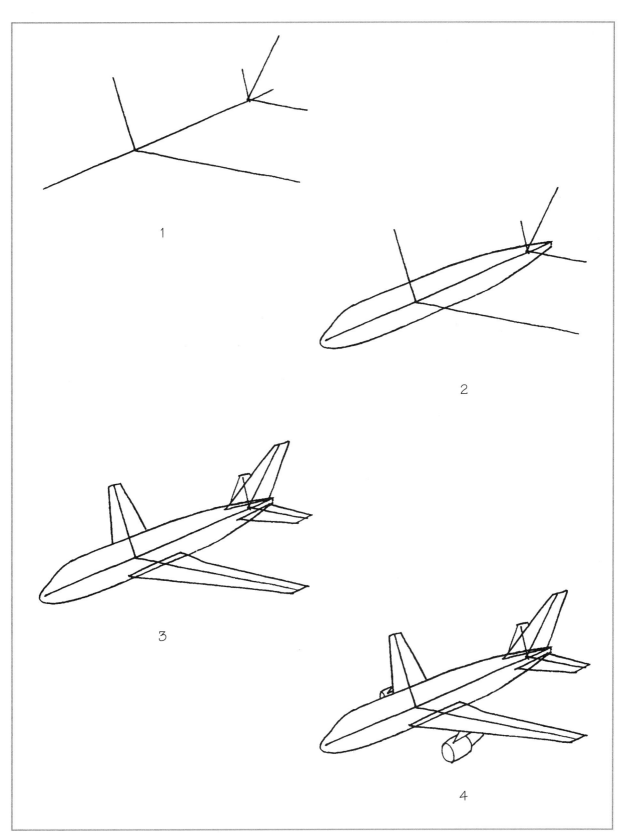

1

2

3

4

Illustration 2.10—Sketching Technique

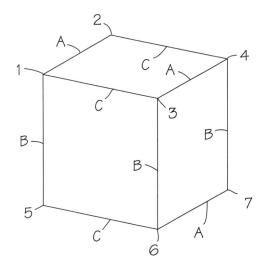

Note that:

Lines A are about 30 degrees above horizontal and parallel to each other.

Lines B are vertical and parallel to each other.

Lines C are about 10 degrees below horizontal and parallel to each other.

Point 1 is positioned about midway between points 3 and 4 along the vertical direction.

Point 2 is positioned about midway between points 1 and 3 along the horizontal direction.

Point 4 is positioned about midway between points 1 and 2 along the vertical direction.

Point 6 is positioned about midway between points 2 and 4 along the horizontal direction.

Illustration 2.11—Studying Objects for Clues

Since patent drawings are not art, but technical illustrations, your drawings do not have to be artistically perfect or beautiful. However, they do have to be reasonably accurate in depicting the structure of the invention.

3. Page Layout

Each figure (drawing) should be big enough to show all of its details clearly. If several different figures are still small enough, they should be placed on the same sheet of paper to avoid using too many sheets of paper. If you know the drawings will fit on the same sheet, but you have difficulty positioning them during the sketching process, you can draw them on different sheets, cut them out, paste them on one sheet, and then make a photocopy. If the figures are too large to fit on one sheet, just use as many additional sheets as necessary.

4. Ink in the Lines

After the sketch is complete, trace over the pencil lines with ink. You should always use rulers and templates to guide the ink pen, and avoid drawing anything freehand as much as possible. When you are sure the ink is completely dry (the ink is still wet if it is shiny), carefully erase any stray pencil marks that remain visible outside the ink lines. If you are impatient and start erasing too soon, the eraser will smear the wet ink lines.

5. Trace Over the Sketch if Necessary

A sketch is sometimes so rough that its lines are each formed by a jumble of pencil marks, so that the exact placement of the final ink line is unclear, as shown in Illustration 2.12. Instead of risking a permanent mistake by inking over a rough sketch, you can put a fresh piece of paper on it, and trace the sketch carefully with a pencil on a light box. The tracing should be much clearer, so that you can apply ink lines on it with confidence.

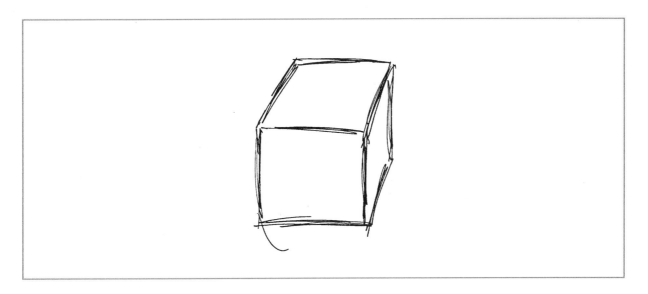

Illustration 2.12—Rough Sketch Makes Final Line Placement Unclear

E. Drawing to Scale

Doing the drawings is much easier if you have some idea about the dimensions of the object you wish to draw, because such dimensions will enable to you to draw the object with the correct proportions. If you have the object available, you can use a ruler to measure all of the critical dimensions, and a protractor to measure the angles of angled lines and surfaces. But then again, if the object is available, tracing it is easier. If you don't have the object, you can create some estimated measurements.

1. Use Metric Measurements if Possible

If you can immediately tell whether 2 $^5/_{16}$" or 2 $^{11}/_{32}$" is longer, you are very familiar with the U.S./English measurement system (inches, feet, etc.). Otherwise, you should use metric units, such as millimeters (mm), centimeters (cm) and meters (m), which are based on multiples of ten. For example, 10 mm equals 1 cm (centimeter), 100 cm equals 1 m (meter), etc.

With the metric system, it is immediately apparent to anyone that, for example, 6.24 cm is longer than 6.22 cm. Also, the millimeter, which is about the thickness of a dime, is a fine enough measurement for most tasks. The centimeter is equal to 0.39 inch, which is roughly equal to the diameter of a typical pen. As long as you remember visually how long the millimeter and the centimeter are, adapting yourself to work with the metric system is easy.

2. Converting Dimensions With a Scale Rule

Drawing to scale is easy with a scale rule. For example, if an object is 46 cm (18.1") tall and 20 cm (7.9") wide, you can use a 1:2 scale rule to draw it at half its actual size, so that it is about 23 cm (9") tall and 10 cm (4") wide to fit on a sheet of letter-size paper.

3. Converting Dimensions Manually

If there is no suitable scale rule for an object—that is, if you need to reduce or enlarge it at some odd scale, you can convert the dimensions manually. For example, if an object is 30 cm (11.8") tall, and you want to scale it down to about 23 cm (9") tall to fit on a sheet of letter-size paper, you would divide 23 by 30 to get a reduction ratio of 0.77, then multiply all other actual dimensions by 0.77 to get the scaled down dimensions of the drawing. Conversely, if an object is only 3.4 cm (1.3") tall, you can enlarge it to make a clearer drawing. You can scale it up to, say, 20 cm (7.9") tall by dividing 20 by 3.4 to get an enlargement ratio of 5.88, then multiply all other actual dimensions by 5.88 to obtain the drawing dimensions.

Use the following steps to scale dimensions manually:

- Determine the maximum height and width of the object to be drawn. For example, 20 cm tall and 30 cm wide.
- Determine the maximum height and width of the desired drawing area. For example, 23 cm tall and 18 cm wide.
- Determine the scaling factor. In this example, although the object is not as tall as the drawing area, it is wider than the width of the drawing area, so it has to be reduced to fit. Therefore, divide the width of the drawing area (18 cm) by the width of the object (30 cm) to obtain 0.60, which is the scaling factor.
- Multiply all the actual dimensions of the object by the scaling factor to obtain the scaled dimensions. In this example, multiply all the actual dimensions of the object by 0.60 to obtain the scaled dimensions. For example, multiply 20 cm (height of object)

by 0.60 to obtain 12 cm as the height of the drawing, multiply 30 cm (width of object) by 0.60 to obtain 18 cm as the width of the drawing, and multiply all other dimensions accordingly.

4. Drawing

If the object is just the right size to fit on a sheet of paper, it may be drawn at actual size, that is, a scale of 1:1. Refer to Section F, below, for more instructions on drawing. If your drawing is to a different scale, after you have scaled all the important dimensions, draw the object by drawing each line with the scaled dimension, as shown in Illustration 2.13, which illustrates a vacuum sander.

F. Drawing Different View Angles

You may choose to illustrate an object with only orthogonal views (front, side, top, etc.), because these are relatively easy to do. However, if they do not illustrate the object clearly enough, you should also draw one or more perspective views, which may be optionally drawn with foreshortening if desired.

1. Orthogonal Views

In an orthogonal view, the object is drawn as if one side of it is parallel to the drawing surface. A tapered block is used to illustrate the point in Illustration 2.14. Imagine positioning a sheet of paper parallel to the front side of the block, and projecting an image of the block onto the paper. Any surface that is parallel to the paper appears at full size. For example, the size of the front side's image is the same as that of the actual block's, that is, height "B" of the image is the same as height "B" of the actual block. Orthogonal views typically

include many vertical or horizontal lines that represent the sides, top, and bottom of an object.

2. Angled Lines in Orthogonal Views

Surfaces that are angled relative to the drawing sheet do not appear at full size: the more angled the surface is to the drawing sheet, the smaller its image appears. For example, the slanted surface of the block is at a large angle relative to the drawing sheet. It is quite long on the actual block, but its image on the drawing sheet appears much shorter: the size "A" of the slanted surface on the drawing sheet is the same as height "A" on the actual block.

Surfaces that are at a right angle to the drawing sheet appear to have zero height, that is, they appear as single lines. For example, the bottom surface of the block is precisely at a right angle to the drawing sheet, so that it appears as a single horizontal line on the drawing sheet. The same goes for the sides and top of the block.

3. Perspective Views

Illustration 2.15 shows a box and a cylinder as viewed from different angles to illustrate what objects in perspective look like. Note that the views taken from the same height as the objects are not as clear as those that are taken above or below. Also note that the sides of the objects appear at full height when seen at the same level, but they appear shorter when the objects are seen from slightly below or slightly above, and even shorter when the objects are seen from almost directly below or almost directly above. The inverse is true for the tops and bottoms, which are not visible (having a zero height) when seen at the same level, but they appear larger when the objects are seen from slightly below or slightly above, and even larger when the objects are seen from almost directly below or almost directly above.

Actual Object

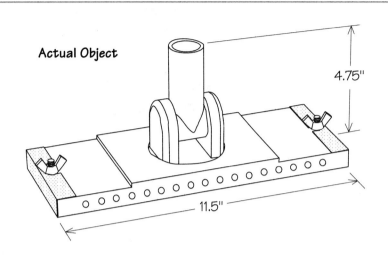

Simple calculations converting dimensions:

Actual Dimensions x Scale Factor = Drawing Dimensions

| 11.5" | 0.6 | 6.9" |
| 4.75" | 0.6 | 2.85" |

Scale Drawing

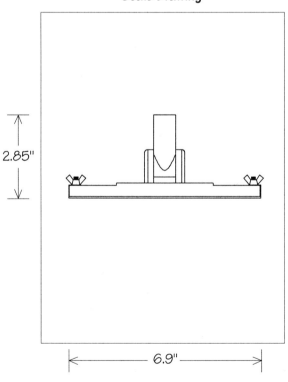

Illustration 2.13—Scale Drawing

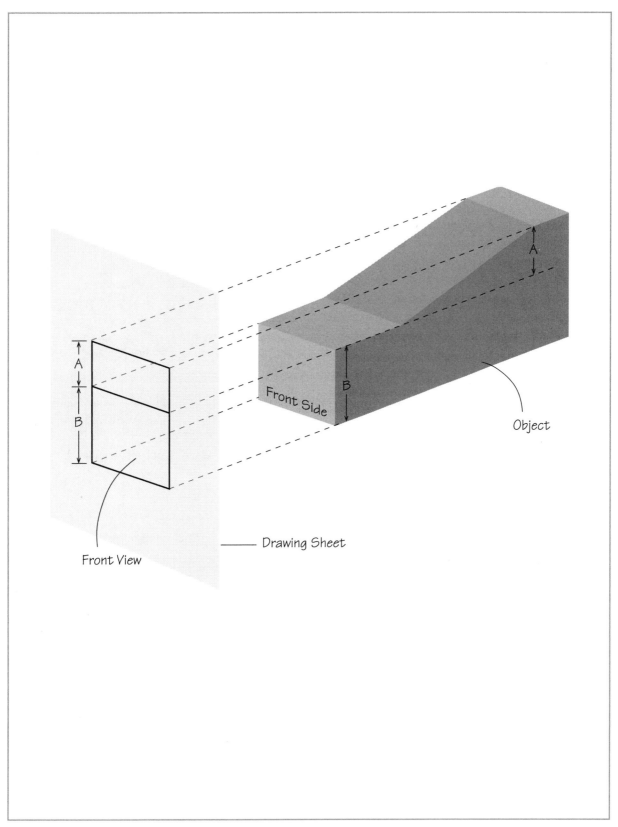

Illustration 2.14—Angled Lines in Orthogonal View

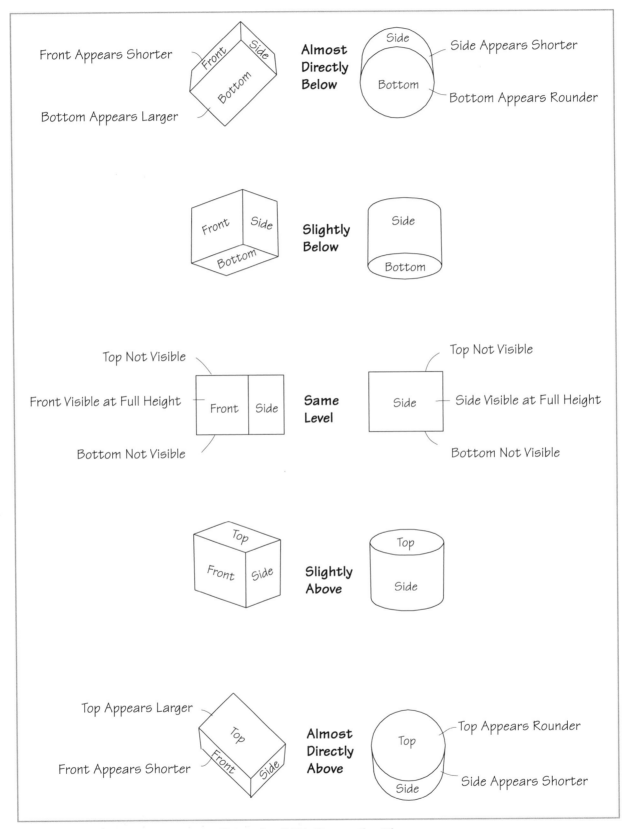

Illustration 2.15—Perspective Views

It is not possible to show in this book what every possible shape would look like from different angles. Illustration 2.15 serves as a rough guide to help you image how your drawing should look. If you still have trouble, refer to Section D, above, for information on studying similar objects for clues.

4. Circles in Perspective

Circles appear as ellipses (a squashed circle) in perspective views. An ellipse has a "major axis" (the line extending across the widest part of the ellipse) and a "minor axis" (the line extending across the narrowest part of the ellipse). The drawing of a cylindrical rod and a hole on a box are illustrated in Illustration 2.16. The dashed lines are all temporary lines that should be sketched lightly in pencil and erased later.

Cylinder:

Step 1. Draw the parallel sides of the cylinder.
Step 2. Sketch the major axis between the ends of the sides. The major axis is the same as the diameter of the cylinder.
Step 3. Sketch the minor axis perpendicular to and through the midpoint of the major axis. The more perpendicular the cylinder is to the drawing surface (the more the end of the cylinder is seen straight-on), the longer the minor axis; whereas the more oblique (angled) the cylinder is to the drawing surface (the more the end of the cylinder is angled away), the shorter the minor axis. The length of the minor axis is difficult to determine exactly, so simply use a length that makes the ellipse look "about right."
Step 4. Draw the ellipse so that it is symmetrical about the major axis, and symmetrical about the minor axis.

Hole on a Box:

Step 1. Determine the center of the hole, and draw the centerline, which is perpendicular to the front side of the box, and parallel to the horizontal edges of the box.
Step 2. Sketch the major axis through the center and perpendicular to the centerline. The midpoint of the major axis should coincide with the center of the hole.
Step 3. Sketch the minor axis perpendicular to and through the midpoint of the major axis. The more perpendicular the hole is to the drawing surface (the more the hole is seen straight-on), the longer the minor axis; and the more oblique the hole is to the drawing surface (the more the hole is seen angled away), the shorter the minor axis. Again, the length of the minor axis is difficult to determine exactly, so simply use a length that makes the ellipse look "about right."
Step 4. Draw the ellipse so that it is symmetrical about the major axis, and symmetrical about the minor axis.

Sketching an accurate ellipse is difficult to do freehand, so you should use an ellipse template that most closely matches the desired shape. If you have an ellipsograph (possibly discontinued; see Section A2, above), you can draw a more precise ellipse.

5. Translating Views by Plotting

A method for translating an orthogonal view into a perspective view comprises covering the orthogonal view with a grid, then plotting and connecting points on another grid drawn in perspective, as shown in Illustration 2.17. The finer the grid, the higher the plotting accuracy. The grid may be applied on the orthogonal view in pencil, or by covering it with a transparent grid overlay. The grid for the perspective view may be lightly drawn in pencil, so that it may be erased later.

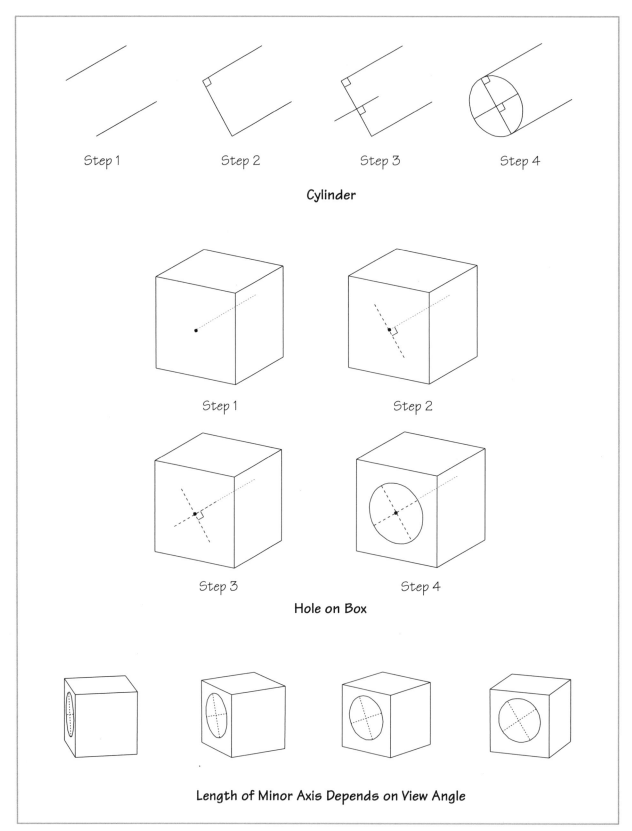

Step 1   Step 2   Step 3   Step 4

Cylinder

Step 1   Step 2

Step 3   Step 4

Hole on Box

Length of Minor Axis Depends on View Angle

Illustration 2.16—Circles in Perspective

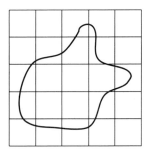

Step 1: Position Grid on Object

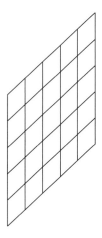

Step 2: Sketch Grid in Perspective

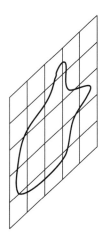

Step 3: Plot Object on Grid

Illustration 2.17—Translating Orthogonal View into Perspective View by Plotting

6. Approximated Perspective Views

A perspective view may be easily approximated by basing it on an orthogonal view, as shown in Illustration 2.18. The perspective view is done by simply drawing oblique (slanted) lines from the orthogonal view, which may be a front, back, top, bottom, or side view. This is akin to making an extrusion of the object, like squeezing toothpaste out of a tube. The oblique lines should be relatively short—much shorter than their actual dimensions—otherwise the perspective view will appear distorted. Such approximated views are not truly perspective, but they are usually acceptable for patent drawings.

G. Drawing Graphical Symbols

Graphical symbols, such as flowcharts and electrical schematics, are relatively easy to draw. There are many other types of graphical symbols used in patent drawings, including fluid-power, architectural, chemical, genetic, etc. It is beyond the scope of this book to discuss the meaning and use of such symbols. If you are unfamiliar with them, you must refer to the literature in the relevant field for guidance.

When drawing graphical symbols, the only problem is arranging the elements of the drawing so that they fit onto a sheet. Use the following method:

1. Sketch the drawing roughly in pencil first to get an idea of its layout on the sheet.

2. If the position of the drawing needs adjusting, you can trace it onto another sheet in the desired position. See Section C, above, for instructions on tracing.

3. If the sketch grows too large to fit onto a single sheet, make another sketch by packing the elements closer together. If that is not possible, or if it makes the elements too crowded, the drawing may be spread over multiple sheets. See Chapter 6, Section C3, for instructions on making multiple page drawings.

4. When you have a satisfactory rough sketch, carefully sketch in the details, using suitable guides to ensure that they are aligned properly. There are templates available for a variety of different symbols, so use them whenever possible. They will make drawing much faster and more accurate.

5. Ink in the lines, and erase any stray pencil marks that remain visible after the ink is completely dry.

H. Practice, Practice, Practice

Remember that if you have never done high-quality drawings before, you should learn and practice basic drawing techniques before starting on your patent drawing project. If you follow this simple and obvious bit of advice, you will ultimately achieve much better results. Again, refer to the books listed at the beginning of this chapter for basic drawing techniques.

Step 1 Step 2 Step 3

Illustration 2.18—Approximated Perspective View

Drawing With a Computer

This chapter provides general instructions for making drawings with a computer. Because there are many different types of computers available, each capable of running many different drawing programs, we cannot provide complete instructions on drawing with a computer: you must rely on the user's manual of your drawing program. Nevertheless, since most drawing programs work in similar ways, we can provide you with a good idea of the necessary equipment and software, how they are used, and the impressive results you can achieve.

If you need help with using a computer or software, please call the computer or software vendor. Neither Nolo Press nor the authors can provide such help.

A. Necessary Equipment and Software

A computer includes two distinct elements: equipment (hardware) and software. The equipment includes a processor, a monitor or display screen, a keyboard, a mouse, and other physical components. Software includes an operating system, such as Windows or Mac OS, that controls the basic functioning of the computer, and application programs, such as word processors and drawing programs, that allow you to do useful work.

1. Bitmap Painting Programs

There are two primary types of drawing programs: bitmap and vector. Bitmap programs, also known as paint programs, manipulate bitmapped images, which are made up of individually controllable pixels or dots. Microsoft Paintbrush, which comes with Windows, and Fractal Design Painter, by Fractal Design Corp., are examples of paint programs. Because they are mainly designed for

making artistic renderings, such as abstract art, and editing photographs, paint programs are completely unsuitable for doing patent drawings.

2. Vector Drawing Programs

Vector drawing programs manipulate lines that are defined by spaced-apart points, like connect-the-dots drawings. The lines are editable by moving the points. There are two main types of vector drawing programs: illustration and CAD (computer-aided drafting or design). CorelDraw, Adobe Illustrator and Aldus Freehand are examples of illustration programs. Such programs are primarily designed for artistic color drawings, such as those used in brochures and magazine ads. Their tools (functions) are not designed for making line drawings, such as patent drawings. Although they can be used to do patent drawings, most are not very good at the task. There are some exceptions, such as illustration programs that are designed for making drawings of graphical symbols like flowcharts. In these cases, illustration programs are excellent. An example of a specialized flowcharting program is CorelFlow, by Corel Graphics.

3. CAD Programs

CAD programs are especially designed for making engineering drawings, which are dimensionally accurate line drawings, so they are the type of drawing program that is most suitable for making patent drawings. CAD programs are further divided into 2D (two-dimensional) and 3D (three-dimensional) programs.

a. Two-Dimensional (2D) CAD Programs

Drawings done in a 2D program are simply computerized versions of a paper drawing, so that they do not include depth information. Each view or figure must be drawn separately. Creating a 2D CAD drawing is roughly similar to making a drawing on paper, except different tools are used in the process. However, even a 2D CAD program has huge advantages over traditional ink drawing techniques: it allows you to construct drawings more easily and with greater precision, and it allows you to edit (modify) them with great ease.

The interface (computer screen display) of a typical 2D CAD program is shown in Illustration 3.1, and includes the following important elements:

1. The workspace or drawing area
2. Tool buttons that may be clicked to perform operations
3. A movable cursor for setting points, pressing tool buttons, selecting drawing objects, etc. The cursor is controlled by a mouse.

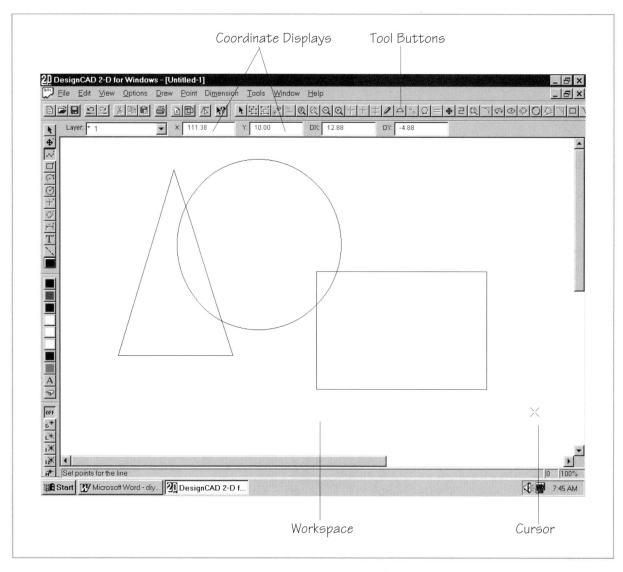

Illustration 3.1—2D CAD Interface

4. Coordinate displays that show the position of the cursor in terms of its horizontal (X) position and vertical (Y) position relative to a reference point. The cursor may be moved a specific distance by dragging it until the desired coordinates are displayed, or by typing the desired X and Y positions in the coordinate displays.

Some 2D programs are listed later in this chapter, in Section 5, below.

b. Three-Dimensional (3D) CAD Programs

Drawings done in 3D CAD programs are three-dimensional representations of actual objects. Although the drawing can only be visualized in two dimensions on the monitor screen and printed output, the drawing file includes information that describes an object in three dimensions. Because creating a 3D drawing is equivalent to building a model of your invention in cyberspace (computer-generated world), such drawings are also called 3D models. A 3D CAD model can be rotated for viewing from any angle, sliced apart to show interior parts, and disassembled, just like the real thing. Thus the main advantage of 3D drawings is that, by making only one 3D drawing of your invention, you can get a variety of 2D views out of it with great ease, and print them as patent drawings.

The interface of a typical 3D CAD program is shown in Illustration 3.2, and includes the following important elements:

1. The workspace, which is the drawing area. This particular program shown has four views of the workspace, the main view on the right being a perspective view, and the three views on the left being (from top to bottom) front, top, and side views.
2. Tool buttons that may be clicked to perform operations.
3. A cursor for setting points, pressing tool buttons, selecting drawing objects, etc. The

different views clearly show the exact position of the cursor in relation to the drawing object, so that additional objects may be created at precise positions with respect to the object.

4. Coordinate displays that show the position of the cursor in terms of its horizontal (X) position and vertical (Y) position relative to a reference point, as well as its depth (Z) position in front of or behind the reference point. The cursor may be moved a specific distance by dragging it until the desired coordinates are displayed, or by typing the desired X, Y, and Z positions in the coordinate displays.

c. 2D Necessary for Doing 3D

While most low-cost CAD programs are 2D only, some are 3D only, and some are both. A 3D-only program lacks the editing tools necessary to turn drawings into final form, so it must be used in conjunction with a 2D program. In that case, the 3D drawing must be converted into 2D format and imported (loaded) into a 2D program for editing, which is inconvenient and may cause compatibility problems in some situations—particularly if the programs are from different publishers. Therefore, it is best to use a program with both 2D and 3D capabilities. Nevertheless, there are good programs that are either 2D only or 3D only. If you like a program that is 3D only, it is best to get a 2D program from the same publisher to ensure compatibility.

4. Charting and Schematic Programs

Most illustration and CAD programs, in addition to being able to make line drawings of objects, are also capable of making flowcharts, electronic schematics and other drawings of graphical symbols.

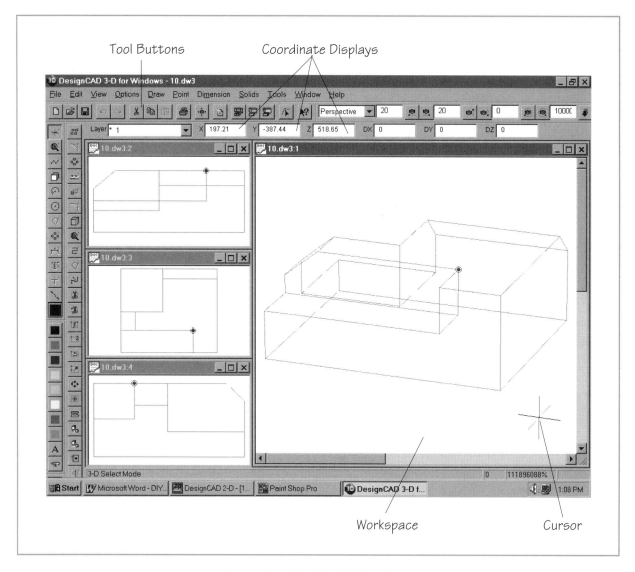

Illustration 3.2—3D CAD Interface

There are also specialized programs that are even more capable of making drawings of such charts, schematics, and symbols. If you want to draw graphical symbols as well as physical objects, you may try a CAD program first to see if it is adequate. If you are only interested in drawing graphical symbols, or if you plan on making many such drawings, you may consider using a program specialized for such symbols.

5. List of Software

The following low- to mid-range priced programs are perfectly suitable for making patent drawings of all but the most complicated inventions. These programs are available directly from their publishers, who sell them at full price, but most are also available from retailers and mail order catalogs for much less. The prices listed below are approximate retail prices, not the publishers' list prices.

2D/3D Programs

TurboCAD 2D/3D

Type: 2D/3D CAD
Requires: Windows 3.1 or Windows 95
Price: about $80
Publisher: Imsisoft (Tel: 800-833-8082;
           Internet: http://www.imsisoft.com)
Notes: The most inexpensive program with both 2D
           and 3D capabilities

CorelCAD

Type: 2D/3D CAD
Requires: Windows 95
Price: $585 ($235 competitive upgrade)
Publisher: Corel Corp. (Tel: 613-728-0826;
           Internet: http://www.corel.com)
Notes: Produced by a respected software publisher

CADKEY

Type: 2D/3D CAD
Requires: DOS, Windows 3.1, or Windows 95
Price: about $700
Publisher: CADKEY Inc. (Tel: 203-298-8888;
           Internet: http://www.cadkey.com)
Notes: Free demo and fully functional DOS Light
           version available

MiniCad 6

Type: 2D/3D CAD
Requires: Windows 3.1, Windows 95, or Mac
Price: $600
Publisher: Graphsoft (Tel: 800-486-8053;
           Internet: http://www.graphsoft.com)
Notes: Free demo available; 30 day money back
           guarantee

Adobe Dimensions

Type: 2D/3D
Requires: Mac System 7
Price: $200
Publisher: Adobe Systems (Tel: 415-961-4400)
Notes: For Macs only

2D-Only Programs

TurboCAD

Type: 2D CAD
Requires: Windows 3.1 or Windows 95
Price: about $35
Publisher: Imsisoft (Tel: 800-833-8082;
           Internet: http://www.imsisoft.com)
Notes: Free demo available

Corel Visual CADD

Type: 2D CAD
Requires: Windows 3.1 or Windows 95
Price: $350
Publisher: Corel Corp. (Tel: 613-728-0826;
           Internet: http://www.corel.com)
Notes: Produced by a respected software publisher

DesignCAD 2D

Type: 2D CAD
Requires: DOS, Windows 3.1, or Windows 95
Price: $160
Publisher: ViaGraphics (Tel: 918-825-4844)
Notes: Free demo available

3D-Only Program

DesignCAD 3D

Type: 3D CAD
Requires: DOS, Windows 3.1, or Windows 95
Price: $240
Publisher: ViaGraphics (Tel: 918-825-4844)
Notes: Free demo available

As a very rough general rule, the more expensive programs are more capable, but there are exceptions. Most of the inexpensive programs are adequate for doing simple to moderately complicated drawings, such as Illustrations 6.22 to 6.29 in

Chapter 6, which were first created with a 3D program, then edited with a 2D program. As with any software, each of these programs has its strengths, weaknesses, and even bugs. To see which one is right for you, ask for brochures and demos from the publishers, and read reviews in computer magazines. Some demos, or demonstration programs, are automatic presentations that show you some of the program's features, while others are limited functional versions that you can actually use. You can also seek the advice and opinion of experienced users through the discussion groups on online services such as CompuServe and America Online, and on Internet news groups, such as alt.cad. In these groups, people can post e-mail-style questions, receive answers, and discuss various topics of interest.

⚠ *The above list of software is provided for your convenience only; Nolo Press and the authors do not endorse any of these products.*

6. Computers

Of course, to use the above programs, you will need a computer. For many years, Macintosh computers had the edge on ease-of-use, but PCs (IBM-compatibles) running Windows 95 are now just as easy to set up and use. PCs also have a significant advantage in price and the sheer volume of available software, so they will be the focus of the following discussion.

If you already have a PC, it may or may not be suitable. CAD programs are relatively demanding software—that is, they typically require moderate to high levels of processing power to work acceptably. Therefore, PCs with 286 or lower processors are not powerful enough for running them. PCs with 386 processors are barely adequate for running some low-end DOS and Windows CAD programs. PCs with 486 processors, and particularly Pentium processors, can run Windows CAD programs quite well. These are generally much easier to use than DOS CAD programs. Each program has different

hardware requirements, which are listed in its brochure and on its box. However, the stated minimum requirements are typically too optimistic —that is, they describe a machine that will run the program, but would be far too slow to run it satisfactorily. The following are specific recommendations for two different computers at different budget levels.

Computer Comparison

A realistic bare minimum machine:

- 486DX2-66 processor
- 8 MB* RAM memory
- 540 MB* hard disk
- Accelerated video board with 1 MB* DRAM
- 14" monitor with at least 800 x 600 resolution, no more than a 0.28 dot pitch, and a refresh rate of at least 70 Hz at 800 x 600 (These specifications define a good monitor)
- Keyboard
- Three button mouse
- Windows 3.11

Cost: roughly $1,000.

A mid-level machine that will perform very well:

- Pentium 100 processor
- 16 MB* RAM memory
- 1 GB** hard disk
- Accelerated video board with 2 MB* DRAM
- 17" monitor with 1024 x 768 or higher resolution, no more than a 0.28 dot pitch, and a refresh rate of at least 70 Hz at 1024 x 768 (These specifications define a good monitor)
- Keyboard
- Three button mouse
- Windows 3.11 or 95

Cost: roughly $2,000.

* MB = megabyte

** GB = gigabyte

Computers with similar configurations are available from a great number of vendors, who typically allow customers to order machines with a custom combination of parts. The specs listed are valid as of this writing, but considering the speed at which the computer industry advances, they may be completely obsolete by the time you read this. Of course, the more a computer exceeds these minimum requirements in terms of processor speed and amount of memory, the better it will run your software. Windows 95 software is listed here because an operating system is typically pre-installed on new computers. Note that Windows 3.11 performs acceptably on 8 MB of RAM memory, and very well on 16 MB, but Windows 95 really requires at least 16 MB to perform acceptably, and 32 MB to really fly. Running Windows with less than the recommended memory will result in very slow performance.

Tips on Buying a New Computer

1. All computers become obsolete in a few years, so getting the fastest one available will ensure that you will get the longest useful life out of it.

2. A computer with mid-level performance will generally give you the most bang-for-the-buck.

3. Most PC vendors, whether big or tiny, do not really manufacture their computers. Instead, they purchase most or all of the snap-together components from the same pool of component manufacturers and simply assemble them. Therefore, there is generally no correlation between name/price and performance/reliability. However, there may be a difference in technical support—that is, how helpful they are in answering your technical questions after you buy a computer.

4. Read computer magazines, such as PC Magazine, for the most up-to-date equipment and software recommendations, and rankings of PC vendors on technical support performance.

5. Seek the advice of experienced users through online services such as CompuServe and American Online, or through news groups on the Internet, such as alt.cad.

6. Computer superstores offer the largest selections of products. Consumer electronics stores also sell computers, but their selections are relatively limited. "Mom-and-Pop" computer retailers with low overhead generally offer the best prices. If you prefer to buy by mail, ask for catalogs from the advertisers in computer magazines. Computers by name-brand vendors, such as Compaq, Dell, and Gateway, are safe choices, but they tend to be much more expensive than those by hole-in-the-wall, "mom-and-pop" retailers.

7. Printers

A printer is necessary for putting the drawing on paper. Do not use dot matrix printers; they do not produce output acceptable to the PTO. Ink jet printers may make satisfactory patent drawings on special ink jet paper that prevents the ink from feathering (becoming fuzzy by bleeding between the fibers of the paper). Most available ink jets print in color. However, the least expensive color ink jets do not have a separate black ink cartridge, so they can produce only a simulated black that is really a dark green. Only color ink jets that use a separate black ink cartridge can produce truly black output, which is required by the PTO. Drawings produced by ink jets, even those with a separate black cartridge, are sometimes objected to by the PTO as not being black enough, sharp enough, etc. This problem can be avoided by submitting a photocopy of the printer output. Cost for a typical ink jet printer: roughly $200.

Low-cost laser printers can produce the best patent drawings for the money. The lowest cost lasers produce output at 300 dpi (dots per inch), but there are slightly more expensive ones that produce output at 600 dpi. The PTO sometimes even objects to drawings produced by 300 dpi lasers for having lines that are not smooth enough. The PTO can be quite picky! 300 dpi laser printers are usually good enough, but go for the 600 dpi printers if you can spare the additional expense. Cost for a typical 300 dpi laser printer: roughly $400. Cost for a low-end 600 dpi printer: roughly $400.

8. Peripherals

You can also get optional peripherals (add-on components) to expand your capabilities or make drawing easier. Such peripherals include:

1. A scanner (256 level gray scale [different levels of gray only] or color) for scanning photographic prints of your invention for tracing in a CAD program. Scanners are available in hand-held and flatbed versions, the latter being better but more expensive. Specific recommendations: Logitech Scanman (handheld), and HP 4C (flatbed). Cost: roughly $70 to $150 for handhelds, and $400 to $900 for flatbeds.

2. A fax machine and a fax modem. A fax machine can be used as a scanner by faxing a photograph or drawing to a fax modem in a computer. Two separate lines or phone numbers are necessary to do this. There are also adapters (about $100) that will allow a fax machine to fax to a fax modem without having two phone lines. However, most fax machines do not approach stand-alone scanners in scan quality. A fax modem is often available as standard equipment in most new computers. Specific recommendations: Brother Intellifax (fax machine), and US Robotics Sportster (fax modem). Cost: roughly $300 for fax machines and $120 for fax modems.

3. A digital camera with 640 x 480 or higher resolution for taking digital photographs that can be transferred directly into a computer, without the need for a scanner. A zoom lens is a must. Specific recommendations: Chinon ES-3000 and Kodak DC50. Cost: roughly $650 to $900.

4. A digitizer tablet with a pen input device for freehand sketching. A digitizer is a flat plate that translates the position of an input device on top of it into cursor position on screen. Most digitizer tablets come with a puck or mouse-shaped input device, but some come with a pen input device, which is a must for freehand sketching. Specific recommendation: Wacom Artpad II. Cost: roughly $130.

Creating a Triangle

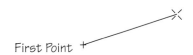

First Point

1. Set First Point to Define First Corner

Second Point

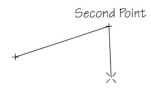

2. Set Second Point to Define Second Corner

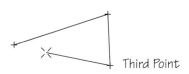

Third Point

3. Set Third Point to Define Third Corner

Fourth Point

4. Set Fourth Point at Same Position as First Point to Close Lines

Creating an Ellipse

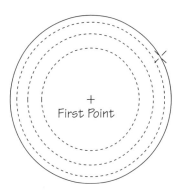

First Point

1. Set First Point to Define Center

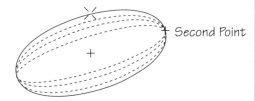

Second Point

2. Set Second Point to Define Major (Long) Axis

Third Point

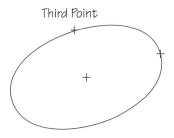

3. Set Third Point to Define Minor (Short) Axis

Illustration 3.3—Creating an Object

Distorting an Object

1. Select the Object by Dragging a "Selection Box" Around It

2. Object Is Surrounded by a "Bounding Box" With Square "Handles"

3. Cursor Changes Shape When Positioned on a Handle

4. Drag Handle to New Position

5. Object Is Distorted

Editing Lines

1. Cut Part of Object With Cut Off Box

2. Object Is Cut

3. Display Points on Line by Using Special Cursor

4. Drag Point to New Position

5. Line Is Repositioned

Illustration 3.4—Editing Objects

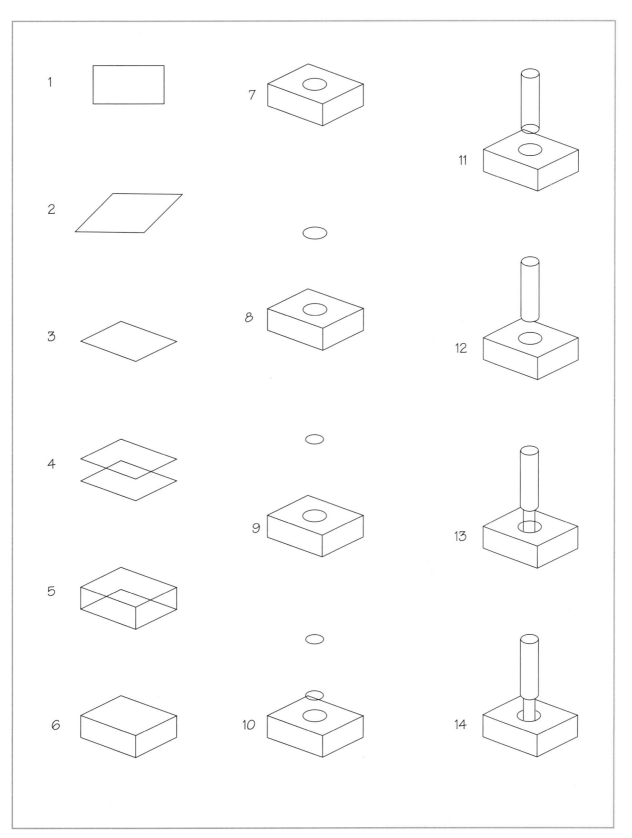

Illustration 3.5—Drawing With 2D CAD

Step 7: Draw an ellipse with the Ellipse tool for the hole on top of the box. Refer to Chapter 2, Section F, for details on drawing circles in perspective.

Step 8: Copy the ellipse and paste it above the original along the same centerline, to create the top of the handle.

Step 9: Slightly reduce the size of the handle top with the Scale tool.

Step 10: Copy the handle top, and paste it below the original to make the lower end of the handle.

Step 11: Draw the sides of the handle by selecting the Line tool and drawing two lines.

Step 12: Delete the excess portion of the lower end, and the portion of the base behind the handle with the Trim tool.

Step 13: Draw the sides of the connecting rod by selecting the Line tool and drawing two lines.

Step 14: Delete the portion of the hole and box behind the connecting rod with the Trim tool.

The joystick was simply drawn until it "looked about right," without regards to dimensions, because it would have been difficult to determine the dimensions of its lines in perspective. Refer to Chapter 2, Section F, Illustration 2.20, for a discussion of the appearance of objects in perspective.

3. Study Similar Objects for Clues

During or after the creation of a drawing, you can easily edit it to achieve the desired result. If you have trouble making the drawing look right, study an object that is shaped similarly to the one you are drawing, or the portion of the object you are drawing. For example, if you have trouble drawing a box in perspective, find a similar box, position it in the same view angle that you are drawing, and study its lines to get an idea of what they should look like, as shown in Illustration 3.6.

When studying the lines, look for qualities such as their angles relative to the vertical or horizontal axes, angles relative to other lines on the same object, and relative length of each line with respect to other lines on the same object, etc. If you are careful in making these observations, you should be able to create satisfactory drawings. However, if you are still having difficulty, you may consider making a model and tracing it, as described in Section C, below, or use a 3D program to create an accurate 3D model, as described in Section D, below.

Patent drawings are not art, but rather technical illustrations, so they do not have to be artistically perfect or beautiful. However, they do have to be reasonably accurate in depicting the structure of the invention. They must also be reasonably well-executed—that is, they must be neat, not sloppy or rough.

4. Page Layout

Each figure (drawing) should be big enough to show all of its details clearly. If several different figures are still small enough, they should be placed on the same sheet of paper to avoid using too many sheets of paper. Page layout is easily accomplished by simply selecting and dragging the figures to desired positions on the page, as shown in Illustration 3.7. Although paper is cheap, and printing many sheets is easy, using fewer sheets increases paper handling convenience for you, and reduces the amount of page flipping the examiner must do. However, if the figures are too large to fit on one sheet comfortably, use as many additional sheets as necessary; do not crowd the figures.

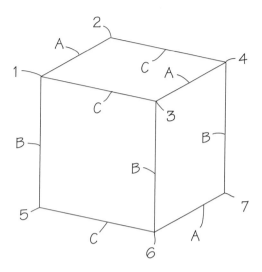

Note that:

Lines A are about 30 degrees above horizontal and parallel to each other.

Lines B are vertical and parallel to each other.

Lines C are about 10 degrees below horizontal and parallel to each other.

Point 1 is positioned about midway between points 3 and 4 along the vertical direction.

Point 2 is positioned about midway between points 1 and 3 along the horizontal direction.

Point 4 is positioned about midway between points 1 and 2 along the vertical direction.

Point 6 is positioned about midway between points 2 and 4 along the horizontal direction.

Illustration 3.6—Studying Objects for Clues

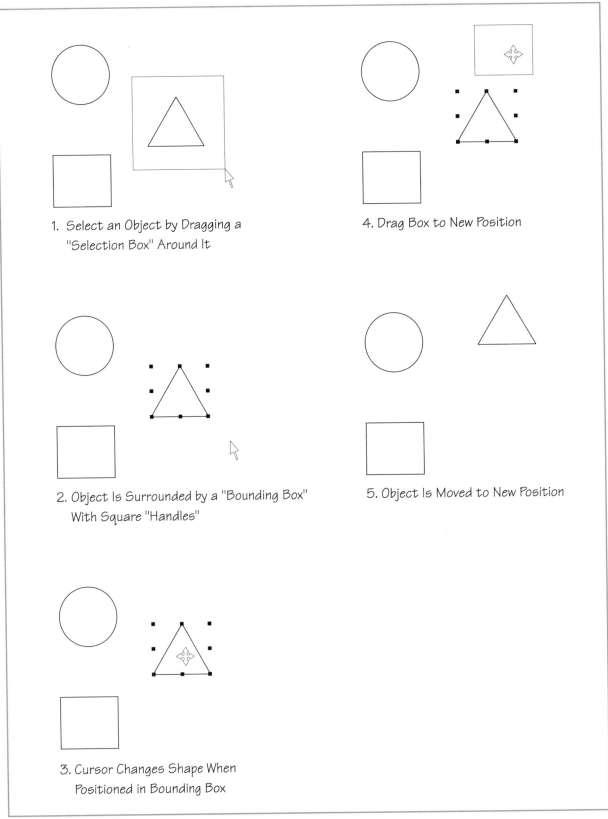

1. Select an Object by Dragging a "Selection Box" Around It

2. Object Is Surrounded by a "Bounding Box" With Square "Handles"

3. Cursor Changes Shape When Positioned in Bounding Box

4. Drag Box to New Position

5. Object Is Moved to New Position

Illustration 3.7—Repositioning an Object

General Drawing Tips

1. Use different line widths for different drawing elements. For example, use 0.2 mm or 0.3 mm lines for the object, and 0.1 mm or 0.15 mm lines for lead lines and hatching. Refer to Chapter 6, Section H, and Chapter 7, Section J, for more information on line types and widths.

2. Use different colors for different line types, so they can be easily distinguished. For example, red for 0.1 mm solid lines, green for 0.2 mm solid lines, yellow for 0.2 mm dashed lines, blue for 0.5 mm solid lines, cyan for hatching, etc.

3. Separate the colors into different layers, so each layer contains lines of the same type and width that can be easily changed by selecting the entire layer and assigning a new type or width to all its lines.

4. Select the font (lettering style) before applying the reference numbers. (See Chapter 8, Section D, for details on reference numbers.) This is because if you change the font after the numbers are applied, they will shift position, and will no longer line up properly with the lead lines. (See Chapter 8, Section E, for details on lead lines.) The amount of shifting depends on the fonts used before and after the change.

C. Making Drawings by Tracing Photos

A very easy way to make a realistic drawing is to trace a photograph of an actual object, such as the prototype of an invention. (Refer to Chapter 4 for tips on how to take good pictures.) The procedure for tracing a photo with a CAD program is generally as follows:

1. Acquire an image of the object by either taking a picture and scanning a photographic print, or by taking a picture with a digital camera. Scanners and digital cameras come with their own software for acquiring and handling images. In either case, the picture is saved as an image file on a computer.

2. Open or load the image with a 2D CAD program. Each program can load specific types of graphics files: BMP, PCX, TIFF, etc. Therefore, when you save the file after scanning or taking a picture with a digital camera, you must save it in a format that is usable to your CAD program.

3. Select the Line tool to trace straight lines, or the Curve tool to trace curves, as shown in Illustration 3.8. Zoom in as necessary to see small details better.

4. When you finish tracing, you can either set the image to be invisible and unprintable, or delete it to save disk space, so that only the tracing is visible and printable.

5. Edit the drawing to fine tune the positioning of the lines and to clean up extraneous lines.

Tracing Tips

1. Put the tracing and the image on separate layers to prevent them from interfering with each other. Depending upon the program you use, there may be certain procedures that must be followed to ensure that the tracing stays on top of the image instead of being covered by the image.

2. Use thick lines, such as 0.2 mm to 0.3 mm, for the outline and major features of the object; and thin lines, such as 0.1 mm, for rounded edges and other curved contours.

3. Lines representing rounded edges should be broken, and their ends should not touch the outlines of the object. See Illustration 3.8.

1. Trace Straight Lines With Line Tool

3. Complete Tracing

2. Trace Curved Lines With Curve Tool

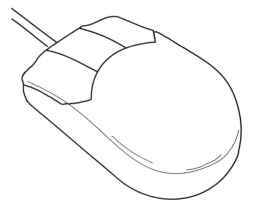

4. Delete Image to Leave Line Drawing

Illustration 3.8—Tracing Photograph of Object

1. Digital Camera Photos Are Blurry

Digital cameras, other than extremely expensive professional models ($5,000-$30,000), have maximum resolutions that are only about 640 x 480 pixels. Such a resolution is enough for a simple object, but it is too low to provide a detailed picture of an object with relatively small details. A typical photo taken by a digital camera is shown in Illustration 3.9. The object appears detailed enough, but after zooming in on a small area, it will appear blurry, so that the exact shape of details will be hard to understand. Therefore, it cannot serve as a guide for precise line placement, which means you may have to estimate the true shape of some of the smaller details. Nevertheless, a digital camera can usually provide an image that is traceable with a little effort. Tracing a blurry image is better than not having one to trace at all.

2. Scanning a Photographic Print

A film camera, even a cheap one, can produce photographic prints that are sharper than even the most expensive professional digital cameras. Therefore, the sharpest image for tracing can be produced by taking a picture with a film camera, making a 4" x 6" print, and scanning it with a scanner. A 4" x 6" print scanned at 300 dpi (dots per inch) results in a 1,200 x 1,800 pixel image. Thus the scanned image may be much larger and sharper than the image produced by a consumer-grade digital camera, so that it is much easier to trace. However, the larger the image size, the more computing power is needed to process it with reasonable speed. For example, the minimum machine listed above will struggle to handle a 1,200 x 1,800 image. Of course, the print can be scanned at a lower resolution to reduce the image size, but the clarity of the details will be compromised.

D. Drawing With 3D CAD

Three-dimensional (3D) modeling is where CAD really shines. As already discussed above, in Section 3, unlike a 2D drawing, a 3D model is a three-dimensional representation of an object in cyberspace (computer space). Such a model is represented by a wire frame, which defines the edges and surfaces of the object. The model may be rotated and zoomed in or out for viewing from any angle or distance desired. 2D "snapshots" may be taken from any angle or distance to create different drawing figures. Therefore, by making just one 3D model, many different 2D drawings may be produced with ease.

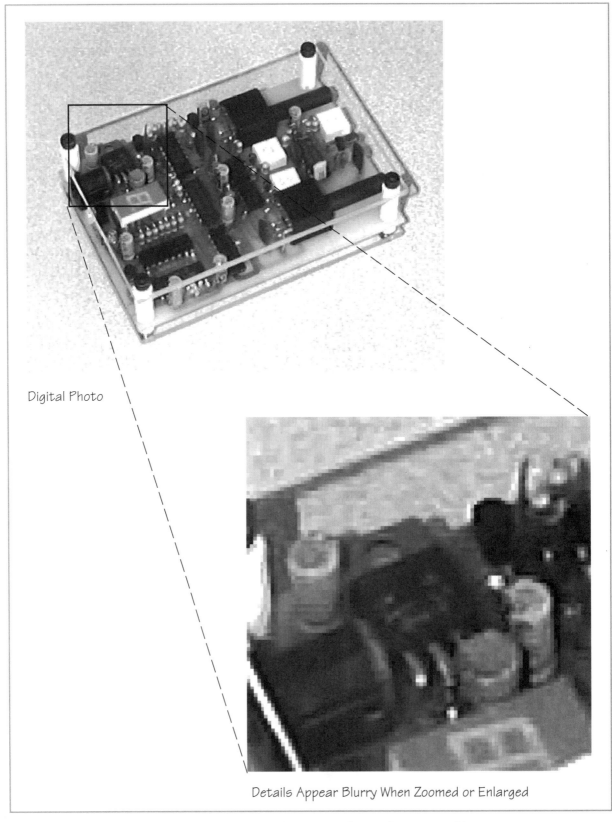

Digital Photo

Details Appear Blurry When Zoomed or Enlarged

Illustration 3.9—Digital Photos Are Blurry When Zoomed in

1. Available Tools and Functions

A typical 3D CAD program includes shape-creating tools, such as Box, Cylinder, Sphere, Cone, Line, Curve, etc. The entities created by these tools may be modified with editing tools, such as Extrude, Sweep, Drill, Lathe, Slice, Cut Off, Chamfer, Trim, etc.

For example, as shown in Illustration 3.10, a box may be created by selecting the Box tool. After setting the first point to define one corner, dragging the cursor creates an adjustable 2D box. Moving the cursor to a position behind the first point and setting the second point defines the diagonally opposite corner, and creates the 3D box. Likewise, a sphere can be created by selecting the Sphere tool. After setting the first point to define the center, dragging the cursor creates an adjustable rough outline of the sphere. Setting the second point defines the radius and creates the sphere. Although these objects are treated by the program as solids, they are normally shown on screen as see-through wireframes.

Other standard and non-standard shapes can also be easily created. As shown in Illustration 3.10, one way to create an irregular shape is by drawing a line or curve, and modifying it with tools such as Extrude and Sweep. A great variety of 3D models can be created by combining and modifying different shapes and lines. The complexity of the models you can build is limited only by your imagination and the available tools in the program.

When two or more objects are combined to create more complex shapes, they must be accurately positioned relative to each other. This is achieved by watching the coordinate displays (Illustration 3.2) on the screen to ensure that the points are set in the proper positions. Various functions are also available for ensuring accuracy in point setting and object positioning. Some of such functions include Snap-Grid, which confines the cursor to movements of a preset distance, such as 1 mm, and Snap-To, which causes a selected entity to snap or stick to a selected position on another object.

2. Making a Drawing in 3D

Just as there are many different ways to construct an actual object, there are many different ways to make a 3D model, even when using the same program. These construction methods will become apparent after you have studied the program's manual and gone through the exercises in the tutorial. As an example, the joystick, first drawn in 2D in Section B, above, is created in 3D in Illustration 3.11 with the following steps:

Step 1: Make the base. Select the Box tool, set one point to define one corner, and set another point a desired distance way to define diagonally opposite corners of the box.

Step 2: Make a cylinder that extends into the base. Select the Cylinder tool, set one point at the center of the box to define the center of the cylinder, set a second point a desired horizontal distance way to define the radius of the cylinder, and set a third point a desired vertical distance away to define the height of the cylinder.

Step 3: Select the Subtract tool, and select the cylinder and the base. The cylinder will be automatically "subtracted" from the base to create a hole.

Step 4: Select the Cylinder tool, set a point at the bottom of the hole to define the center of the rod, and set two more points at desired positions to define the radius and height of the rod.

Step 5: Select the Line tool, and draw a line having desired dimensions above the connecting rod to define half the handle's profile.

Step 6: Select the Fillet tool, and set points on the corners of the line to round them off.

Step 7: Sweep the line to create the handle.

Step 8: Select the Fillet tool, and set points on the edges of the box to round them off.

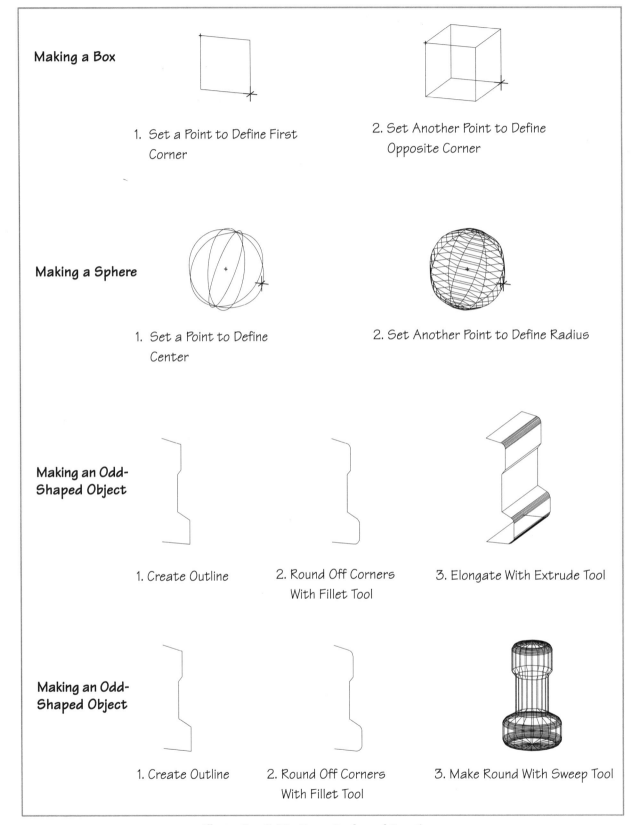

Making a Box

1. Set a Point to Define First Corner

2. Set Another Point to Define Opposite Corner

Making a Sphere

1. Set a Point to Define Center

2. Set Another Point to Define Radius

Making an Odd-Shaped Object

1. Create Outline

2. Round Off Corners With Fillet Tool

3. Elongate With Extrude Tool

Making an Odd-Shaped Object

1. Create Outline

2. Round Off Corners With Fillet Tool

3. Make Round With Sweep Tool

Illustration 3.10—Some Tools and Functions

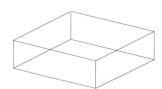

1. Make a Box

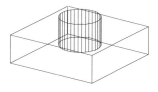

2. Make a Cylinder Extending into the Box

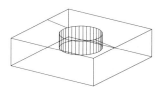

3. Subtract Cylinder From Box to Make Hole

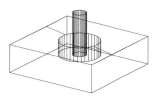

4. Make a Cylinder in Hole

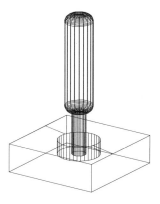

5. Draw Line to Define Half Profile of Handle

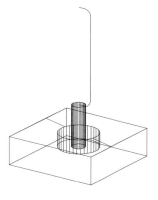

6. Round Corners

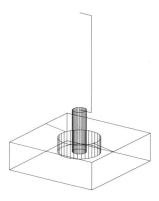

7. Sweep Half Profile to Make Handle

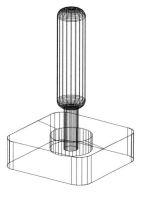

8. Round Edges of Box

Illustration 3.11—Drawing in 3D

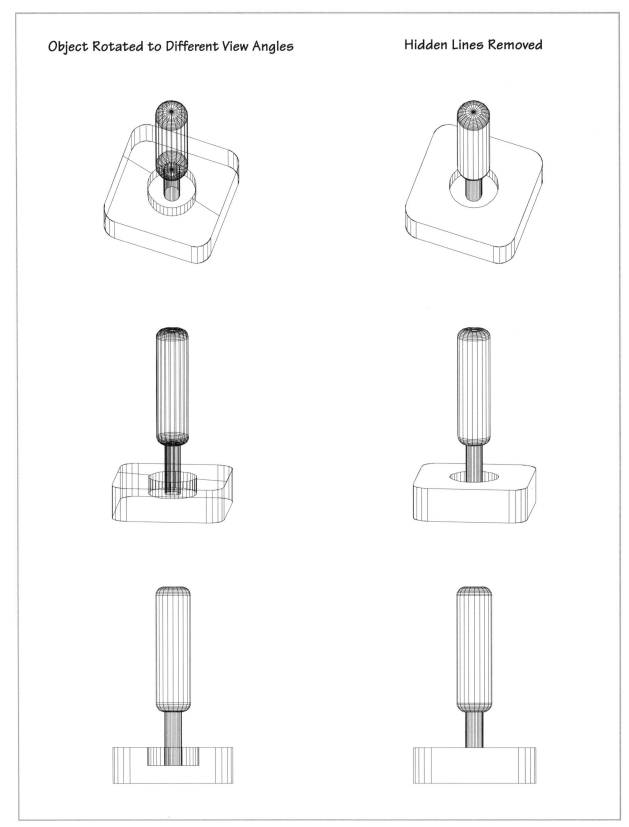

Object Rotated to Different View Angles   Hidden Lines Removed

Illustration 3.12—Rotating Object to Obtain Different 2D Views

The joystick is thus created in 3D, with even a bit more detail than that shown in Section C. The model can be easily rotated and zoomed to produce a number of different views from different angles and distances, as shown in Illustration 3.12. These views can be saved as individual 2D drawings without the hidden lines (lines that should not be visible because they are behind solid surfaces).

3. View Distance

In addition to view angle, the view distance (zoom factor) is typically adjustable in CAD programs. As shown in Illustration 3.13, when the view distance is large, the object appears only slightly distorted. At the other extreme, when the view distance is very short, the object appears very distorted. The most realistic drawing is produced at a moderate view distance. You can easily experiment with the view distance to see what the best setting is for each particular drawing. View distance in CAD is equivalent to the technique of foreshortening in traditional drawing techniques (see Chapter 1, Section B).

4. Cleanup

The conversion from 3D to 2D is not always without problems; 2D drawings are often produced with missing or unwanted lines. For example, 3D programs represent curved surfaces, such as cylinders and spheres, with many flat facets. Some programs provide the option of eliminating the facet lines when the model is converted into 2D, but other programs do not. In that case, the facet lines must be deleted manually. Unless you use a combined 2D/3D program, you must import (load) the 3D drawing into a separate 2D program for editing.

Trace rather than erase. *If a drawing has a large number of facet lines, such as a drawing with many spheres and cylinders, it may be easier to trace the lines you want to keep instead of erasing the unwanted ones. Make the original drawing a single color and put it in one layer. Put the tracing on a separate layer, and make it a different color so you can see your progress. After tracing is finished, delete the original drawing.*

5. Great 2D Drawings for Free

Making a 3D model is different from starting in 2D. However, it is no more difficult than making a single 2D drawing, and provides the significant benefit of allowing you to easily create as many 2D drawings from as many view angles as you wish. Each 2D view is visually accurate.

6. Shading

A 3D model can be used to produce accurately shaded drawings for a design patent application. See Chapter 7, Section H2, for a detailed discussion of the technique.

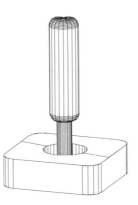

Large View Distance—Object Appears Slightly Distorted

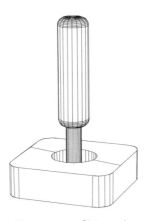

Moderate View Distance—Object Appears Realistc

Very Short View Distance—Object Appears Distorted

Illustration 3.13—Effect of Zoom on Realism

E. Drawing Graphical Symbols

Two-dimensional (2D) CAD programs may be used to make drawings of graphical symbols, such as flowcharts and electronic schematics. Some programs come with a variety of types of ready-made symbols—typically electronics symbols—that can be easily arranged together to create a drawing. There are also add-on symbol libraries from a variety of vendors. If the symbols you need are not available, you can easily make them from scratch. You only need to make each symbol once; after it is created, it may be easily duplicated as many times as necessary.

There are many ways to make a drawing of graphical symbols. As an example, one way of making an electronic schematic is as follows:

1. Open the drawings of the symbols you need (symbols are typically stored separately in different files). Copy and paste them into one drawing.

2. Arrange a few symbols next to each other, and draw the connections between them, as shown in Illustration 3.14.

3. Arrange more symbols next to the first ones, and draw the connections between them. Whenever an identical symbol is needed, simply duplicate one already in the drawing by copying and pasting. The orientation of the symbols may be changed by rotating them with the Rotate tool.

4. Repeat as many times as necessary to gradually build up the drawing.

F. Summary

Making drawings with a computer is relatively easy and fast. Even the most inexpensive drawing programs will enable you to make great-looking drawings without requiring you to have any traditional drawing skills. If you are skilled in traditional drawing techniques, and simply wish for the ability to modify your drawings with ease, then a 2D program will suffice. However, 3D programs, or combined 2D/3D programs, are so much more powerful that they are well worth the additional expense and effort to buy and learn them.

The functional advantage of CAD over pen and rulers is similar to that of a word processor over a typewriter, but pen and rulers still hold a cost advantage. If you already have a computer, or are planning on buying one anyway, we strongly suggest that you use CAD. Otherwise, traditional drawing techniques are still perfectly viable. Remember, most professional patent drafters still use traditional (pen and ink) techniques.

The drawing file should not be erased until the application process is finished, in case changes to the drawings must be made. Also, it is a good idea to back up the file by copying it onto a floppy disc or other backup device, so that if you accidentally overwrite or erase it in the computer, it can be restored by copying it back onto the hard drive.

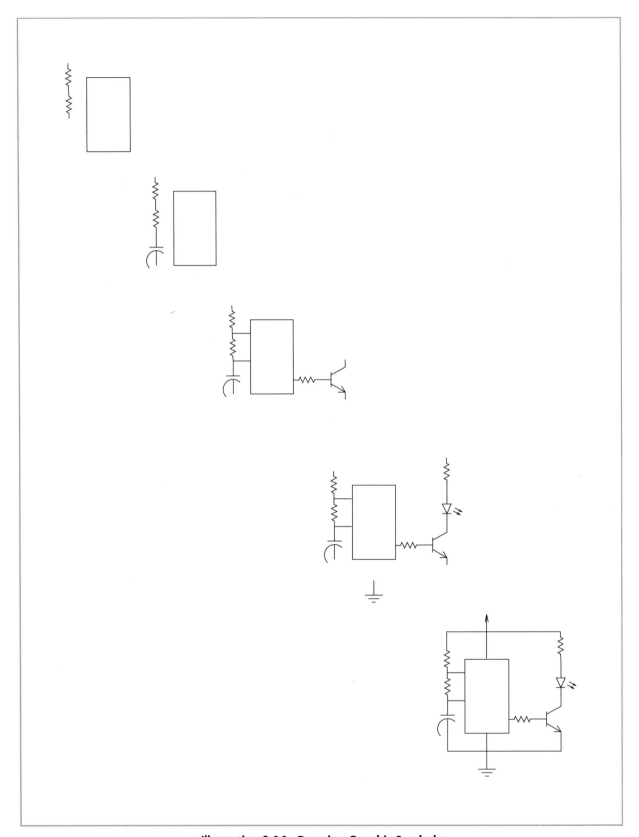

Illustration 3.14—Drawing Graphic Symbols

"Drawing" With a Camera

Color or black-and-white photographs are normally accepted as informal patent drawings. Under new rules promulgated in 1994, the PTO now accepts black-and-white photographs as formal utility and design drawings, and color photographs as formal utility drawings (not design), provided that you meet the requirements discussed in Chapter 8, Section C. See also Chapter 5, Section D, for details on formal and informal drawings. If photographs are submitted as informal drawings, formal line drawings must be submitted to replace them when the application is allowed (approved), unless you successfully petitioned to have the photographs accepted as formal drawings.

If you have built a model or prototype of your invention, and it is in the form that you wish to patent, you may photograph it instead of making line drawings. Also, if your invention cannot be adequately illustrated by line drawings, photography may be the only alternative. This chapter provides instructions on how to take suitable pictures.

A. Advantages and Disadvantages

The advantages of photography include the lack of artistic skills required, the relatively short time needed to take a picture, the ability to illustrate subtle or certain complex images much better than can line drawings. The disadvantages include the need for the novice to acquire a new set of skills, the difficulty of photographing very small objects clearly, and the difficulty or impossibility of photographing internal parts without cutting up an invention.

B. Inventions Suited for Photography

Virtually all inventions that are tangible objects may be illustrated with photographs. Some inventions, such as crystalline structures, textile fabrics, cell structures, metallurgical microstructures, etc., cannot be adequately illustrated with line drawings, so photography would, in fact, be the preferred method. However, taking clear photographs of complicated inventions or inventions with tiny parts may be difficult. Of course, photography cannot be used for illustrating non-tangible inventions and graphical representations, such as schematic diagrams, charts, etc.

A PTO rule prohibits the inclusion of both photographs and line drawings in the same application. Therefore, a tangible invention should not be illustrated with photographs if line drawings are also needed to illustrate its aspects that cannot be photographed, such as a sectional view.

C. Photographs Must Show Invention Clearly

It is vital that the photographs show all the important parts of the invention clearly, without any ambiguity. This is easy to accomplish for very simple inventions or inventions with large parts, but difficult to accomplish for complicated inventions or inventions with tiny parts. Your application may be rejected if the photographs do not illustrate the invention clearly enough. Such a rejection usually can only be overcome by filing new photographs or line drawings to show the invention more clearly in a continuation-in-part (CIP) application. However, the features newly revealed by the clearer photos or line drawings will not get the benefit of the original filing date. (See Chapter 9, Section I, for details on filing CIPs.)

D. Equipment

Photographs submitted as patent drawings must clearly show all the important details. The minimum equipment necessary for taking clear pictures includes the following:

1. A camera with a flash and a zoom lens. Snapshot cameras (roughly defined as cameras with non-removable lenses) without a zoom lens may take acceptable pictures in some situations, depending on the size and shape of the object (the proper photographic term is "subject," but "object" is used here for consistency). (See Section B, below, for a discussion of how flash and zoom are used and why they are necessary.) Consumer-grade digital cameras, which have resolutions of about 640 x 480 pixels, are not suitable because they produce pictures that are less sharp and detailed than even the most inexpensive film cameras. However, such digital cameras are suitable for taking pictures for tracing. See Chapter 3, Section C.

2. Black and white 35 mm or larger print film, so that it can be enlarged to patent drawing size—that is, up to about 8" x 10"—without becoming too grainy. It should have a speed of ISO (also known as ASA) 100 for filming in sunlight, or ISO 200 for filming in medium light with flash. Faster film—that is, film with a higher ASA—should not be used, because the faster it is, the more grainy the prints are. 110 film, which is actually about 17 mm wide, is not suitable because it produces grainy pictures, particularly when enlarged. Black and white film should be used because, with a few exceptions, photographs submitted as patent drawings must be black and white. See Chapter 8, Section C, for details. Color film can be used for producing black and white prints, but the results would not be as good as using black and white film.

3. Two portable lamps for illuminating the object. Desk lamps may suffice.

4. A tripod.

E. Taking Pictures

Although artistic photography is a complicated undertaking, patent photography is not, because its only goal is clear, sharp pictures. Aside from reading the manual of your camera, you can take good patent pictures by following the simple rules and tips discussed in this section. (The same rules and tips also apply to taking pictures with a digital camera for tracing, as discussed in Chapter 3, Section C.)

1. Choice of Background

A high contrast background should be used. For example, if the object you want to photograph is light-colored, the background should be dark colored, and vice versa. The background should also be uniform in appearance, and of a solid color, such as a bare table top or a large piece of cardboard, so that there will be no confusion as to what is the background and what is part of the object.

2. Artificial Lighting

Always shoot (take pictures) in a brightly lit area. If necessary, illuminate the object with two lamps positioned around it, so that all of its details are clearly visible, and to avoid dark shadows. Position the lamps to avoid glare reflecting off the object. Objects of dark colors should be lit with very bright lights to make them more visible.

3. Flash

Most snapshot cameras will usually flash when shooting indoors, unless the room is very brightly lit. The flash will create shadows if the object is positioned very close to the camera (due to the different lines of sight between the lens and the flash to the object). If the camera flashes, position a lamp very close to the side of the object opposite the flash, to reduce the shadow, as shown in Illustration 4.1.

4. Sunlight

Avoid putting the object in direct sunlight, which will cast dark shadows that obscure details, as shown in Illustration 4.2. If direct sunlight cannot be avoided, position yourself between the object and the sun, so that you will photograph the side of the object that is most evenly lit. Also position the object to minimize shadows. If at any time there are dark shadows on the object, use the flash to reduce them. Most snapshot cameras have a Day Flash setting that makes the camera flash even in bright light.

5. Maximize Object Size

Make the object fill the viewfinder frame—that is, make it appear as large as possible—so that all of its details are clear. This is particularly important because the photograph must be enlarged. See Illustration 4.3. If an object or a detail is very small, you must use a zoom of high enough power or a macro lens to make it appear large enough.

6. Zoom In to Reduce Distortion and Preserve Focus

If you are using a snapshot camera, stand far enough away so that you must zoom in to make the object fill the frame. Taking a picture without zooming in—for example, if your camera has no zoom—requires you to stand very close to the object. The short lenses used in most snapshot cameras cause the object to appear distorted when shooting up close, as shown in Illustration 4.4. Standing farther away and zooming in substantially

reduces the distortion. The required distance depends upon the size of the object—the larger the object, the farther away you must be.

Zooming in is also necessary when shooting a small object; otherwise, in order to make it fill the frame, you must stand so close that the camera will not focus. This is because snapshot cameras typically have a minimum focus distance of about 1 meter or 3 feet (refer to the manual of your camera). Most snapshot cameras have a focus indicator light in the viewfinder that alerts you when you are too close (refer to the manual of your camera on how to interpret the light signals). Instead of zooming in, you can use the macro setting (extreme close-up) if your camera has one.

7. Depth of Field

If you use a camera with manually adjustable aperture (F-number or F-stop) and exposure (shutter speed), set the aperture to F-16 or higher to make both the front and back of the object in focus—that is, to increase the "depth of field." Set the exposure or shutter speed control on "A" (automatic). Keep in mind that the higher the F-number, the longer the exposure will be, and the steadier the camera must be to take sharp pictures. Therefore, a tripod is a must. The use of a high F-number is necessary, particularly when the shooting distance is small relative to the size of the object. For example, when shooting a telephone from just 60 cm (2 feet) away, or shooting a car from 2 meters (6 feet) away, the aperture should be set to a high F-number. Snapshot cameras typically have fully automatic aperture and exposure that are not user adjustable.

8. Use a Tripod

To ensure sharp pictures, use a tripod to steady the camera, and press the button slowly so as not to move the camera too much.

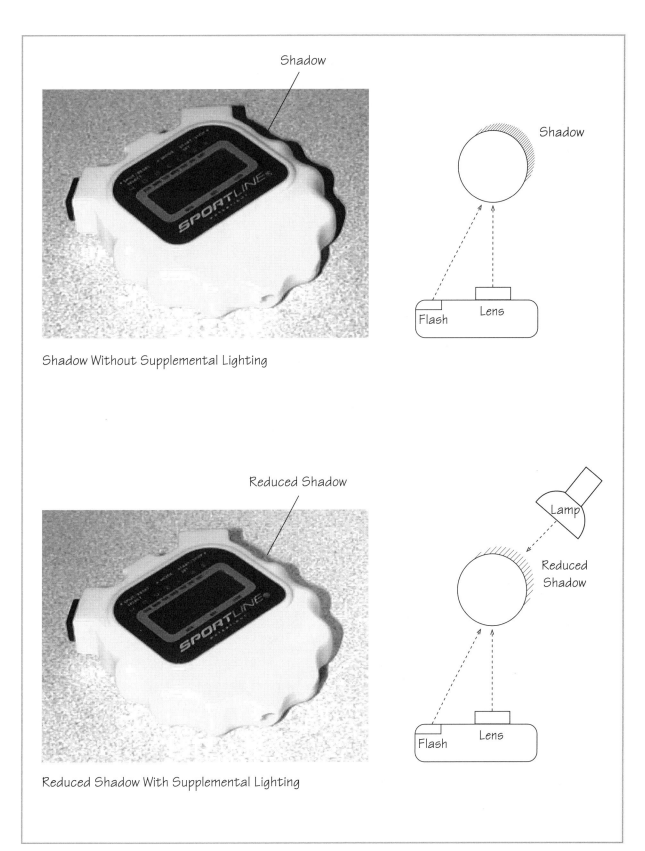

Illustration 4.1—Shadow Reduction With Supplemental Lighting

Dark Shadows

Bright sun without flash

Reduced Shadows

Bright sun with flash

Illustration 4.2—Shadow Reduction in Bright Sun

DON'T make the object appear small in frame

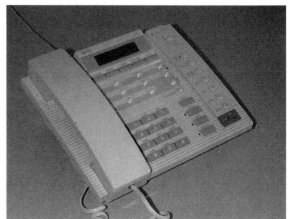

DO make the object fill the frame

Illustration 4.3—Maximize Object Size in Frame

 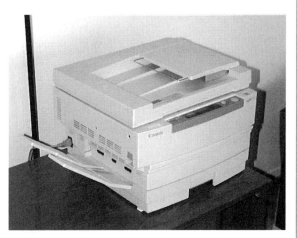

Distorted
Photographed up close without zooming in

Not Distorted
Photographed farther away and zoomed in

Illustration 4.4—Zooming in to Reduce Distortion

9. Take a Lot of Pictures

Photography requires a lot of guesswork. Therefore, you should take a lot of pictures from many different sides, angles, and lighting conditions, so as to increase the likelihood that enough of them will be usable. Write down the settings for each shot, so the best settings may be used again for another roll if necessary.

Black-and-white prints are rare nowadays, so they are about five to six times more expensive than color prints to develop. Therefore, instead of making prints of all the photos, order a contact sheet (a single large print with all the pictures on it in small scale), and select the ones you want to print and enlarge.

10. Developing The Film and Making Prints

There are many places at which you can develop and print color photos, including photo developing

shops, camera stores, and supermarkets. You might have to look around for a photo developer that handles black-and-white film or prints. Typically, photo developing shops provide a full range of services—including black-and-white prints and enlargements to 8" x 10"—services that many supermarkets probably do not offer.

F. Summary

Photographs submitted as patent drawings must clearly show all of the important features of the invention. Taking pictures may be easier than making line drawings, but making the pictures clear enough may require some experimentation. Photography has limitations as a means for producing patent drawings, most notably its inability to make cutaway views (presuming that you don't want to cut up the prototype you made with your sweat and blood). Nevertheless, if your invention is not too small, and you do not need cutaway views, photography is a viable way to make patent drawings. ■

Patent Drawings in General

Federal laws and rules require most patent applications to be filed with drawings. However, there are some situations in which patent drawings are not necessary. This chapter discusses the drawing requirement, when you can get away with not submitting a drawing, and the three types of patent drawings required by the PTO for the various types of patents applied for.

A. The Drawing Requirement

Title 35 of United States Code, Section 113, which is part of the patent laws, states the requirement for patent drawings this way: "The applicant shall furnish a drawing where necessary for the understanding of the subject matter to be patented." Since the vast majority of inventions cannot be clearly conveyed by words alone, drawings are necessarily in almost all patent applications.

1. When Drawings Are Not Required

Drawings may not be required in applications for inventions that can be clearly conveyed by words alone, without any ambiguity. There are very few instances when this can be done, but the following types of inventions are specifically exempt:

1. A process, such as a heat-treating process for hardening metal.
2. Compositions, such as a road surfacing material, a chemical, or a drug.
3. Coated articles or products, such as paper or cloth coated with a water-repellent material.
4. Articles made from a particular material or composition, such as non-slip floor tiles made from recycled tires.
5. Laminated structures, such as plywood.
6. Inventions in which the distinguishing feature is the presence of a particular material, such

as a hydraulic system distinguished solely by the use of a particular hydraulic fluid.

It is impossible to draw some inventions, such as a heat treating process, articles made from a particular material (the material being the only novelty), and compositions. However, if it is possible to illustrate an invention with a drawing, even though it is adequately conveyed by words alone, your examiner may require you to provide one. For example, coated articles and laminated structures are specifically exempt, but since it is possible to illustrate them with drawings that show the different layers of the invention, the examiner may require you to provide the drawings.

Even if your invention is exempt from the drawing requirement, we strongly recommend that you submit drawings whenever possible for the following reasons:

1. Examiners like every patent to contain a drawing, if possible, because it makes the invention much easier to understand.
2. Drawings please the examiner and show him or her that you, the applicant, are making an effort to present the invention as clearly as possible. A pleased examiner is more likely to allow your application.
3. When you get a patent, the drawing will make it easier to recognize as a relevant patent when others do patent searches, so it will be more effective in warning potential infringers, and in preventing others from getting patents on improvements of your invention.
4. It communicates your invention better to any company to which you offer your invention for sale or license and any judge or jury who may ultimately rule on its validity and infringement if you ever have to go to court.

B. If No Drawing Is Submitted With a Patent Application

A patent application is checked for completeness when it is first received at the PTO. If it is filed without a drawing, and the PTO determines that a drawing is necessary for the understanding of the invention, the application will not be given a filing date because it will be considered incomplete. It will also be considered incomplete and not be given a filing date if it mentions a drawing, but none is submitted. In either case, the PTO will notify you to furnish the drawing. When the drawing is received in the PTO, the application will be given a filing date as of the date of the drawing's receipt.

If you have not furnished a drawing because you felt it wasn't necessary for the understanding of the invention, but the examiner requires you to submit one because it is possible to illustrate or explain the invention better with one, then your application will not be denied a filing date. There is no need to explain why you have not filed a drawing, because the examiner will make a decision based solely on the specification (written description of the invention).

> **EXAMPLE:** An inventor files a patent application on a heat-treating process for making softballs that involves specific temperatures and steps. He explains the process clearly and completely in the specification. His application will be assigned a filing date, because no drawing is necessary for the understanding of the invention. However, months after filing, when the application is taken up for examination, the examiner requires the inventor to furnish a flowchart-type drawing for illustrating his heat-treating steps, so that his patent will be an easier-to-understand search reference. The inventor must provide the drawing in the time period set by the examiner—usually three

months—otherwise his application will go abandoned. If the inventor submits the drawing in time, his application will retain its original filing date.

C. Three Types of Patent Drawings

If your invention is not exempt from the drawing requirement, as discussed above, you must provide a drawing. Different types of drawings are required for different types of patents. Refer to *Patent It Yourself* for additional details and help in choosing the proper type of patent to seek for your invention. Let's briefly look at the specific drawings that must accompany each of the types of patents.

1. Utility Patent Drawings

Utility patents cover useful inventions, including:
- Devices, such as tools, machines, electronic circuits, paper clips, etc.
- Methods, such as manufacturing processes, software, surgical procedures, etc.
- Compositions, such as chemicals, drugs, biological material, etc.
- Improvement of known devices, such as an improved vacuum cleaner.
- New uses of old devices, such as using aspirin to speed the growth of pigs.

Each utility patent application must include as many of such drawings as necessary to show every essential feature of the invention, so that its structure and operation can be fully understood. The important elements are labeled with reference numbers, which are referred to in the written description portion of the application. Illustration 5.1 shows a typical utility patent drawing, which happens to be of a geared, rotary scalpel.

Although the example shows a tangible device, utility patent drawings can comprise graphical

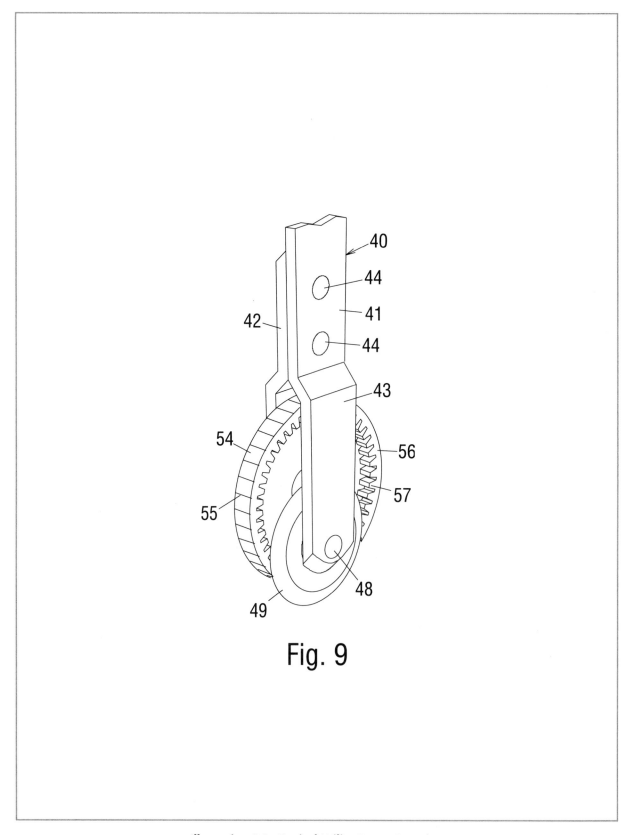

Fig. 9

Illustration 5.1—Typical Utility Patent Drawing

representations, such as flowcharts and electronic schematics, to illustrate abstract concepts. Photographs may be submitted instead of line drawings under certain conditions. (See Chapter 4, and Chapter 8, Section B.) Utility patent drawings are discussed in detail in Chapter 6.

2. Design Patent Drawings

Design patents cover inventions with unique aesthetic qualities. For example, a light fixture design, a shoe design, a camera housing design, or a sofa design. A design patent covers only the aesthetic aspect (styling) of an invention.

An invention may be covered by both utility and design patents: the utility patent may cover its utilitarian aspects, while the design patent may cover its styling.

Illustration 5.2 shows a typical design patent drawing, which happens to be of a floppy disc clock. Each design patent application must include several drawings to show the invention from different viewpoints, so that its appearance can be clearly understood. No reference numbers should be used, because a design patent does not include a description of the invention, other than a descriptive title. The drawings must include adequate shading, that is, lines or marks that depict surface contour. (See Chapter 7, Section H.)

Photographs may be submitted instead of line drawings under certain conditions. (See Chapter 4 and Chapter 8, Section B.) Design patent drawings are discussed in detail in Chapter 7.

3. Plant Patent Drawings

Plant patents cover asexually reproducible plants—that is, those that are reproducible by grafting and cutting, such as flowers. Plants with practical uses, such as medicinal herbs, may also be covered by utility patents. Plants that are novel only in appear-ance, such as ornamental flowers, cannot be covered by utility patents.

Each plant patent application must include as many drawings as necessary to artistically show every distinguishing characteristic of the plant, from as many viewpoints as necessary. If color is a distinguishing characteristic, then color drawings or photographs must be used. (See Chapter 4 and Chapter 8, Section B, for details on photography.) No reference numerals may be used, because a plant patent does not include a description of the plant, other than a descriptive title. Plant patents are extremely rare, so no drawing sample is provided here.

D. Formal and Informal Drawings

The PTO requires drawings to conform to a set of strict standards, discussed in detail in Chapter 8. Drawings that are determined by the PTO's Drafting Branch to comply with the standards are considered to be "formal" drawings, whereas those that do not are considered to be "informal" draw-ings. Any error in a substantially formal drawing will cause it to be considered as "informal." Free-hand sketches are accepted as informal drawings. An application filed with informal drawings will be given a filing date, but the PTO will require formal drawings to be submitted if and when the applica-tion is eventually allowed (approved). However, an application initially submitted with formal drawings will make a better impression on the examiner.

1. Even Informal Drawings Must Be Clear and Detailed

Although freehand sketches can be submitted as informal drawings when the application is filed, they must be detailed and clear enough to fully

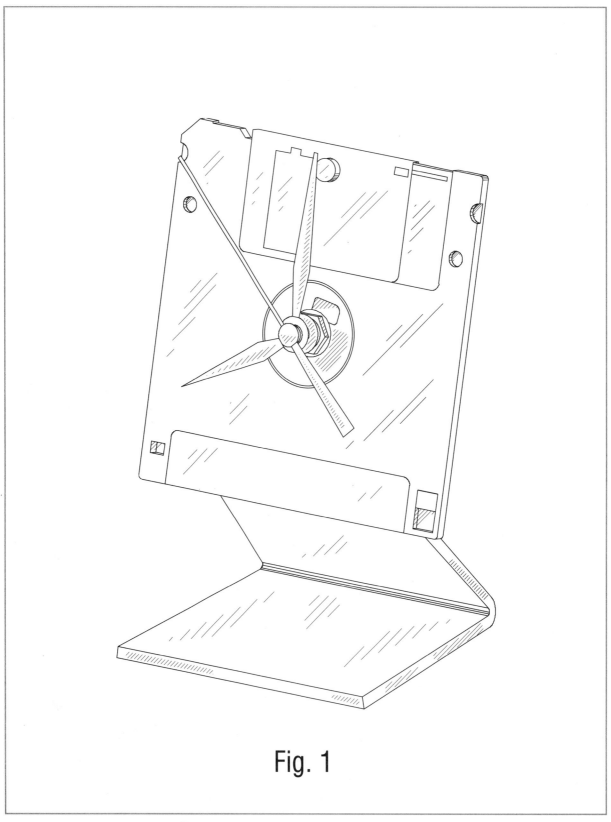

Fig. 1

Illustration 5.2—Typical Design Patent Drawing

convey your invention. The formal drawings that are required after the application is allowed must show the same information as in the informal drawings.

PTO rules prohibit the addition of new matter to the same application. (New matter is any information not in the application as originally filed. New matter can be added by filing a continuation-in-part application, but the new information won't get the benefit of the original filing date. See Chapter 9, Section I, for details.) Therefore, if an examiner objects to a sketch because it does not show a certain part clearly enough to make it understandable, you cannot show it clearer in the formal drawings, because new information (new matter) must be provided to make the part understandable. Consider the difference between blurry and sharp photographs of a newspaper: the text on the newspaper cannot be read in a blurry photograph, whereas it may be easily read in the sharp photograph; so the sharp photograph has more "information" than the blurry one. Therefore, even informal drawings should be very carefully planned and executed, and show all the essential details clearly.

2. Photographs

Photographs, whether black-and-white or color, can be submitted as informal drawings for utility or design patent applications. Black-and-white photos are accepted as formal drawings in utility or design patent applications if they are accompanied by a petition and a petition fee, and if the petition is granted. Color photographs (and color drawings) are accepted as formal drawings in utility patent applications (never design) if they are accompanied by a petition and a petition fee, and if the petition is granted. (See Chapter 8, Section B, for details on such petitions.) Otherwise, formal line drawings must be submitted to replace the photographs when the application is allowed. See also Chapter 4 for details on patent photography, including its advantages and disadvantages.

3. Overview

The following table summarizes the types of drawings or photos that are accepted as informal drawing, formal utility drawing, or formal design drawing.

	Informal Utility or Design Drawing	Formal Utility Drawing	Formal Design Drawing
Sketches	✔		
Black line drawing that does not comply with all PTO rules (see Chapter 8 for rules)	✔		
Black line drawing that complies with all rules (see Chapter 8 for rules)		✔	✔
Black-and-white photo	✔		
Black-and-white photo with granted petition (petition fee required)		✔	✔
Color photo or color drawing	✔		
Color photo or color drawing with granted petition (petition fee required)		✔	

E. Engineering Drawings Are Not Suitable

Inventors who finance the manufacturing and marketing of their inventions usually have a set of engineering drawings or blueprints made. Engineering drawings are created according to engineering standards, which are very different from patent drawing standards. However, they may be submitted as informal drawings, provided they clearly show the invention, and reference numerals are added to label the parts. As discussed above, if such drawings do not show the invention clearly, they cannot be clarified later, so make sure that they are clear and understandable even if they are filed as informal drawings.

If you have computerized engineering drawings, commonly known as CAD (computer aided drafting) drawings, you may be able to modify them into formal patent drawings. See Chapter 3, Drawing With a Computer, for details on CAD drawings. ■

Utility Patent Drawings

This chapter details the specific requirements of formal utility patent drawings; refer to Chapter 5, Section D, for a discussion of formal and informal drawings. Also refer to Chapter 5, Section C, and *Patent It Yourself,* our companion volume, for additional details on utility patents, including the types of inventions that qualify for them and how to prepare the written portion of the patent application.

It is very important to plan the drawings carefully before starting to draw, such as the number of views from different angles, the specific angle of each view, whether to use sectional or exploded views, and how to show the movement of movable parts. The more complicated the invention, the more planning is necessary. There are many issues that must be worked out first. This chapter will help you accomplish that. Therefore, you should understand this chapter before planning the drawings.

A. Amount of Detail Required

The two most important requirements regarding utility patent drawings are:

1. The drawings and the written description (collectively known as the "disclosure") must be detailed and clear enough to show a person skilled in the pertinent field how to make and use your invention. That is, the disclosure must be "enabling." For example, if your invention is a medical laser, your drawings and description must be detailed enough to show a medical laser engineer how to make and use your invention. On the other hand, they do not have to be detailed enough to allow a layman to understand it.

2. The drawings must show all the parts or elements mentioned in the specification, including the description of the invention (technical description) and claims (legal description). For example, if the description and claims mention parts A, B, and C, the drawings must show all three parts.

1. Some Details May Be Left Out

Satisfying the above requirements doesn't mean that the drawings must show every tiny detail. As mentioned, they only have to show enough to enable "a person skilled in the pertinent field to make and use your invention." Therefore, details that such a skilled person would know to provide may be left out. Generally, implicitly-required elements, such as screws that attach parts together, need not be shown or described. If your invention is electromechanical (a mechanical device with electrical parts), you do not have to show the wires if you provide a separate electrical diagram (discussed below in Section F). Some conventional elements may be illustrated with symbolic representations (also discussed below in Section F) if their detailed illustration is not essential for an understanding of the invention. For example, if a conventional, off-the-shelf steam generator is used in a new steam cleaner, the steam generator may be simply represented by a box labeled "Steam Generator."

Also, a conventional element may be mentioned in the description but not shown in the drawings if it is not essential for an understanding of the invention. Such an element should be followed by the note "(not shown)."

EXAMPLE 1: The invention in Illustration 6.1 includes an electrically-driven roller 23 that drives a belt 21r around passive rollers 5 and 6. Instead of showing a detailed drawing of a motor connected to the driven roller, a circle is used to represent it. A simple electrical circuit (the lines and symbols connected to the right of roller 23) shows that the roller is electrically driven by a battery, which need not be numbered or mentioned in the description because it is readily understood by anyone skilled in the field. Note that the rollers are not shown to be connected to any supporting structure, which is acceptable as long as this fact is mentioned in the description. For example, you could write, "The elements shown in Fig.

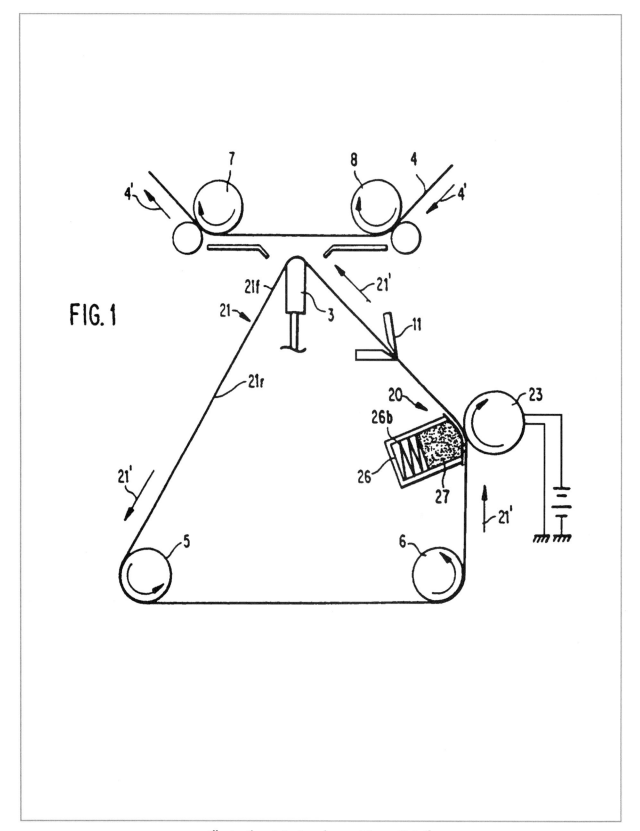

FIG. 1

Illustration 6.1—Leaving out Some Detail

1 are mounted in a suitable housing (not shown)," because the housing would also be readily apparent to anyone skilled in the field.

EXAMPLE 2: The motor vehicle gear position indicator in Illustration 6.2 includes a housing 10 positioned above a steering column 12. Housing 10 need not be shown mounted to any supporting structure, if such structure is conventional, and the description mentions such structure. For example, "Housing 10 is mounted in a conventional instrument panel (not shown)."

2. Better to Include Too Much Rather Than Too Little Detail

There is no precise way to determined exactly how detailed the drawing must be. When in doubt, you should err on the side of providing more detail, because while too much detail would not hurt, too little detail definitely will.

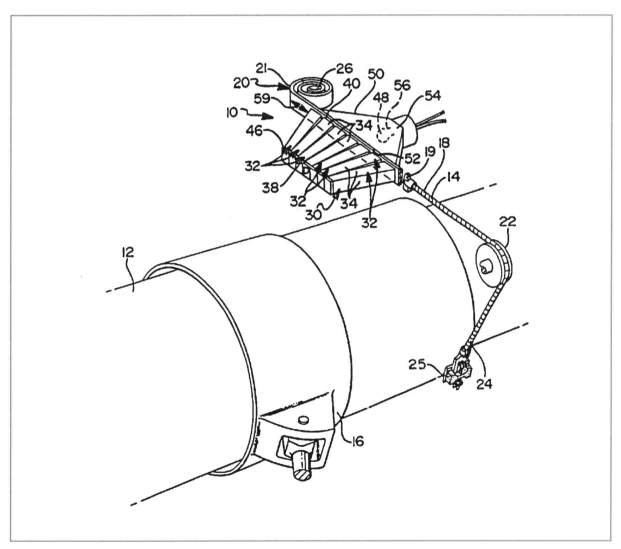

Illustration 6.2—Leaving out Some Detail

3. Dimensioning Is Usually Unnecessary

Do not include the dimensions or angles of the parts in the drawing unless they are crucial to the understanding of the invention, or if they are important in distinguishing your invention from the prior art (similar but older devices). However, if you do include dimensions or angles, the PTO prefers metric units (millimeters, centimeters, meters, etc.) over English units (inches, feet, etc.).

4. Idealized Parts

You can apply for a patent even if you have not made a model. If you have a rough model or prototype of your invention, you do not have to show an exact representation of it in the drawings; you can show it in an idealized form if you wish. For example, if the model has a part that was fabricated by attaching two or more elements together, it can be shown as a single, integral part, such as in Illustration 6.3. Joints between parts that are not meant to be separable—that is, they are

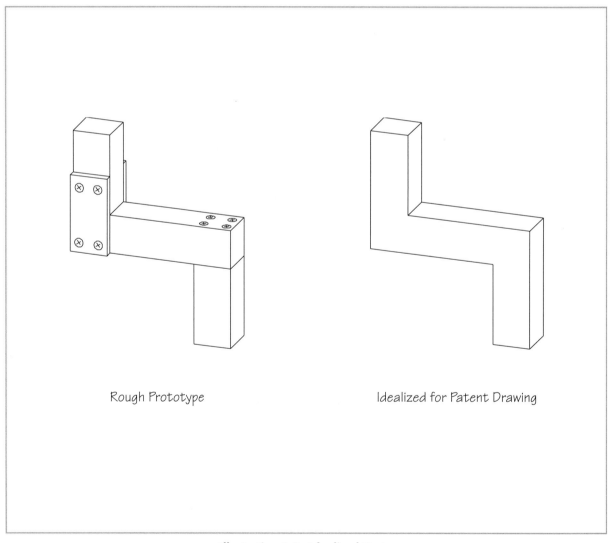

Rough Prototype

Idealized for Patent Drawing

Illustration 6.3—Idealized Parts

permanently connected—can be left out. Also, a detail can be eliminated if it is not necessary for a complete understanding of the invention, and you do not care to claim it.

> **EXAMPLE:** You invented a portable table with a new folding mechanism. When you built your prototype, you used a height adjustable leg that you happened to find in your workshop. You can show the leg in a simplified, nonadjustable form, if:
>
> 1. You feel that the adjustability of the leg is not an important part of your invention;
> 2. You do not describe the leg as being adjustable in the description; and
> 3. You do not describe the leg as being adjustable in the claims.

5. Invention Part of Larger Machine

An invention that forms part of a larger machine, such as a transmission for a motor vehicle, can be shown with just a small portion of the larger machine to illustrate its role. Illustration 6.4 shows a transmission controller that includes a control box 30 and an actuator 28. The dashboard is partially presented to show where the control box is installed, and a "disembodied" transmission 16 is presented to show where the actuator is attached. Such a drawing is perfectly acceptable.

6. Number All Parts

All parts or elements in a utility patent drawing mentioned in the description and claims must be designated with a reference numeral or character. See Chapter 8 for details on reference numerals.

7. Make as Many Drawings as Necessary

Most inventions cannot be clearly understood with one drawing. You must make as many drawing views (separate drawings, also known as figures) as necessary, so that the structure and operation of your invention may be easily understood. Section C details the different types of views that may be used.

8. No Changes Are Permitted After Filing

No "new matter"—that is, technical information not shown or described in the original drawings or description—can be added after the application is filed. For example, the drawings cannot be changed to incorporate substitute parts, new features, or improvements. Therefore, submitting a drawing that does not clearly illustrate the invention is an error that cannot be fixed, because new parts would have to be added to the drawing to improve understandability, and such parts will most likely be considered to be new matter, and therefore would not be accepted by the PTO.

The determination of whether the drawing adequately conveys the invention is made when the application is first examined by an examiner. This usually happens many months to over a year after the filing date. (See Chapter 13 of *Patent It Yourself.*) If the application is rejected because the drawings are not clear enough, the only way to overcome the rejection is to change the drawings and file them with a second application, known as a continuation-in-part application (CIP), which requires a new filing fee. (See Chapter 9 for details on how to overcome rejections.) The new matter filed with a CIP will not get the benefit of the original filing date. (See Chapter 14 of *Patent It Yourself* for more information on CIPs.) Therefore, the drawing is just as important to the whole application as the written portion, and must therefore be clear and adequately detailed.

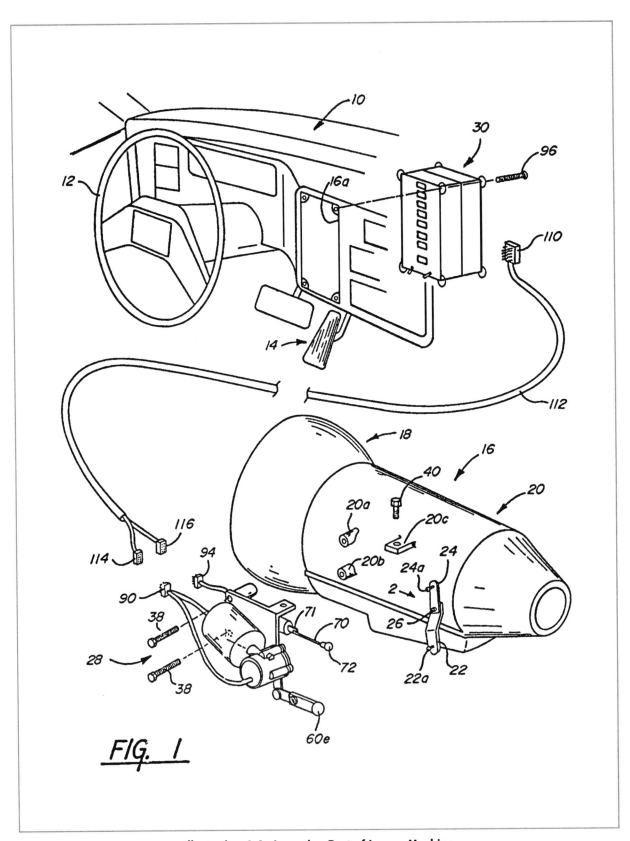

FIG. 1

Illustration 6.4—Invention Part of Larger Machine

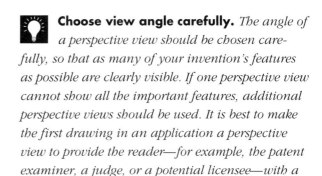

Choose view angle carefully. *The angle of a perspective view should be chosen carefully, so that as many of your invention's features as possible are clearly visible. If one perspective view cannot show all the important features, additional perspective views should be used. It is best to make the first drawing in an application a perspective view to provide the reader—for example, the patent examiner, a judge, or a potential licensee—with a general understanding of the invention.*

How to Check for Adequacy

Having created the invention, an inventor is so familiar with it that he or she sometimes take its structure and operation for granted, and omits small but vital details in the drawing. One way to avoid falling into this trap is to show the drawing and description in confidence to others, preferably those who are skilled in the same field, and ask them if they are clearly understandable. At the very least, carefully consider if they are understandable to someone skilled in the field, but who has never seen your invention before.

B. Types of Views

A complete understanding of an invention often requires multiple drawings showing it from different view angles, sliced open to reveal internal parts, disassembled, etc. Such drawings are called "views." A single sheet of paper may contain several views. The views must not touch each other, and must be far apart enough so that they are clearly separated.

1. Orthogonal and Perspective Views

Orthogonal and perspective views are the mostly commonly used views. (See Chapter 1 for details.) Perspective views are easier to understand than orthogonal views, but they are also more difficult to draw. To save work, you may illustrate your invention with only orthogonal views, but if they are too difficult to understand, you should add one or more perspective views.

2. Exploded Views

The comprehension of some inventions may be enhanced by exploded views, which show the device disassembled. The constituent parts are separated as far as necessary to show the essential details of each one. The moved position of each part is preferably perpendicular to its assembled position, but individual parts may be oriented in any position necessary to show essential details. An exploded view should always be accompanied by another view showing the parts assembled together, as shown in Illustration 6.5. Exploded views are generally unnecessary, because internal parts are usually better illustrated with sectional views (discussed in Section 4, below). If you wish to use exploded views, note the following:

Brackets. As shown in Illustration 6.6, unconnected parts must be "enclosed" by a bracket to show that they belong to the same figure. The bracket can be oriented in any position. The parts in Illustration 6.7 are purposely positioned so that they are all touching to eliminate the need for a bracket, but such positioning is not as clear as that of Illustration 6.6. No bracket is needed for an exploded view, even if the parts are unconnected, if it is the only figure on a sheet. Note that the exploded pipe fittings of Illustration 6.6 and Illustration 6.7 are positioned along a straight line because they share a common axis.

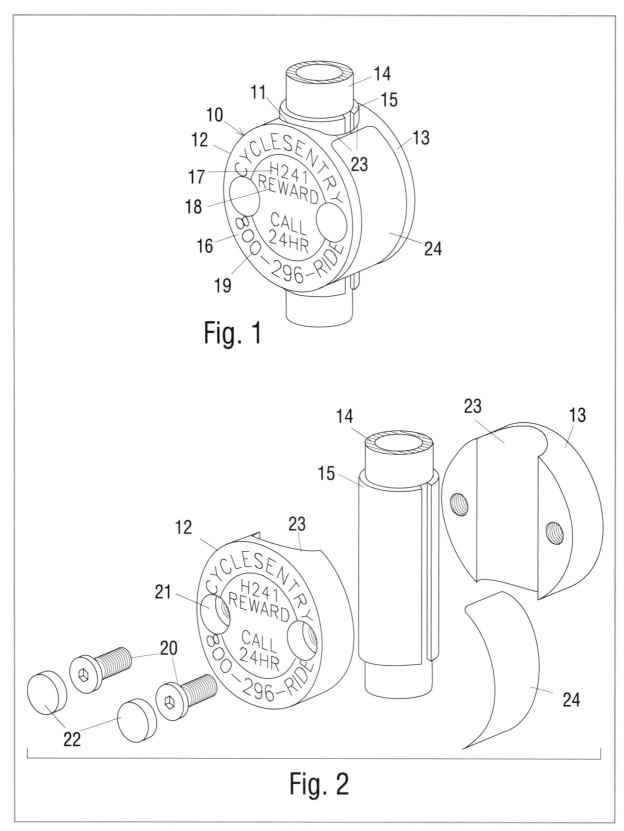

Fig. 1

Fig. 2

Illustration 6.5—Exploded Accompanied by Assembled

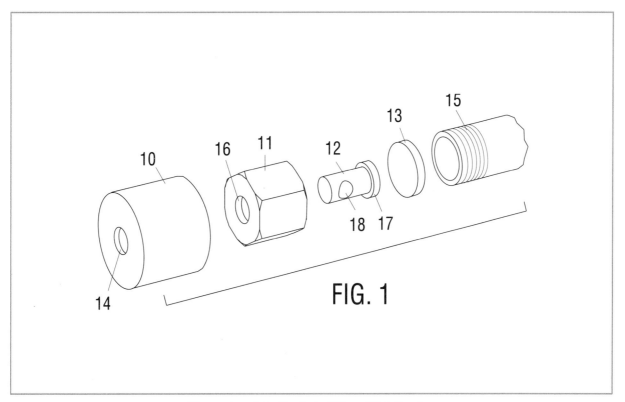

Illustration 6.6—Exploded View With Bracket

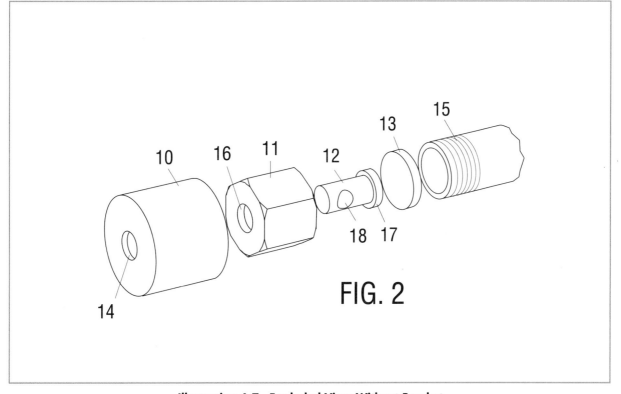

Illustration 6.7—Exploded View Without Bracket

Projection lines. The parts in Illustration 6.8 do not lie along a single axis, so each part is separated as if it is pulled in a straight line from its assembled position. Optional projection lines (the dot-dashed lines) are provided to show how the parts fit together. As shown in Illustration 6.9, when space constraints prevent the parts from being positioned in straight lines from their assembled positions, zigzagged projection lines can be used. No brackets are necessary in Illustration 6.8 and Illustration 6.9 because the projection lines connect all the parts. Projection lines should always be provided when the parts are not separated linearly, so as to make their assembly clear.

3. Partial Views

A very long object that cannot fit onto a single sheet without making the details too small can be broken up, so that each piece can be shown larger than it could have been if the object is shown whole, as long as there is no loss in comprehension. There are several ways to do this:

1. Break it into two or more pieces to fit them onto a single sheet, such as in Illustration 6.10. A simple long bar is used to illustrate the point. The pieces should be connected by projection lines to show how they fit together. If no projection lines are used, a bracket should be used. See Section C2, above, for details on the use of brackets.

2. Shorten it by removing a section, as shown in Illustration 6.11, but no important details should be removed. Again, a simple long bar is used to illustrate the point. It is broken in three places to show the three conventional ways for representing a shortened object. The top type of break is typically for representing a cylindrical object, whereas the other two can be used for anything.

3. Spread it across two or more sheets, such as the drinking bag in Illustration 6.12. The views must be arranged on each sheet so that the sheets can be tiled (positioned like bathroom tiles) next to each other to assemble the complete figure. A dot-dot-dash line should be provided to denote the broken edge of each partial view. Any arrangement of the sheets can be used, for example, side-to-side, top-to-bottom, and rectangular array, as long as the sheets can be assembled without ambiguity or blocking the drawings on each other. Each partial view should be labeled with a figure number having a letter suffix, for example, Fig. 3A and Fig. 3B.

Also, an invention can have a part broken off and omitted if comprehension is not affected. The broken edge is customarily made jagged to indicate its nature, such as the portions of element 14 below its horizontal surface in Illustration 6.13.

4. Sectional Views

If a device has hidden parts or internal features that should be shown, one or more sectional (cutaway) views may be used. Illustration 6.14 shows a general view of a box, Fig. 1, and a sectional view, Fig. 2., of the hollow interior of the box. Whenever a sectional view is used, a broken line with perpendicular end arrows should be placed on a general view, such as in Fig. 1. The broken line between the arrows define the sectioning plane, that is, where the object is sliced. The arrows indicate the view direction. Numbers corresponding to the figure number of the sectional view must be placed next to the arrows, and sized about 5 mm or 1/5" high. If the general view is orthogonal, such as in Illustration 6.14, the position of the sectioning plane is assumed to be perpendicular to the drawing sheet (the sheet of paper the drawing is on).

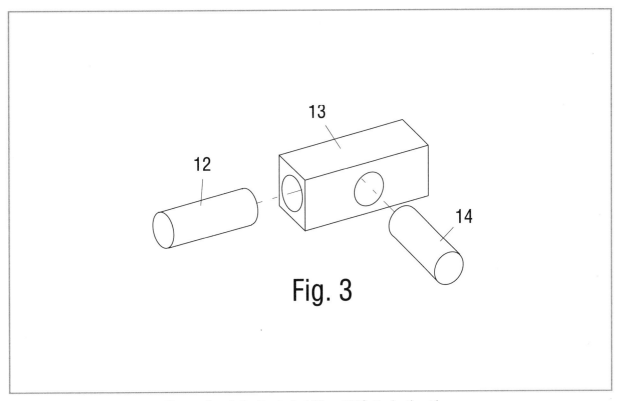

Illustration 6.8—Exploded View With Projection Lines

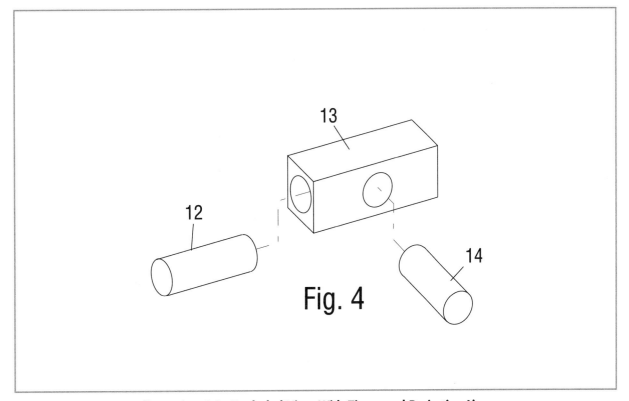

Illustration 6.9—Exploded View With Zigzagged Projection Lines

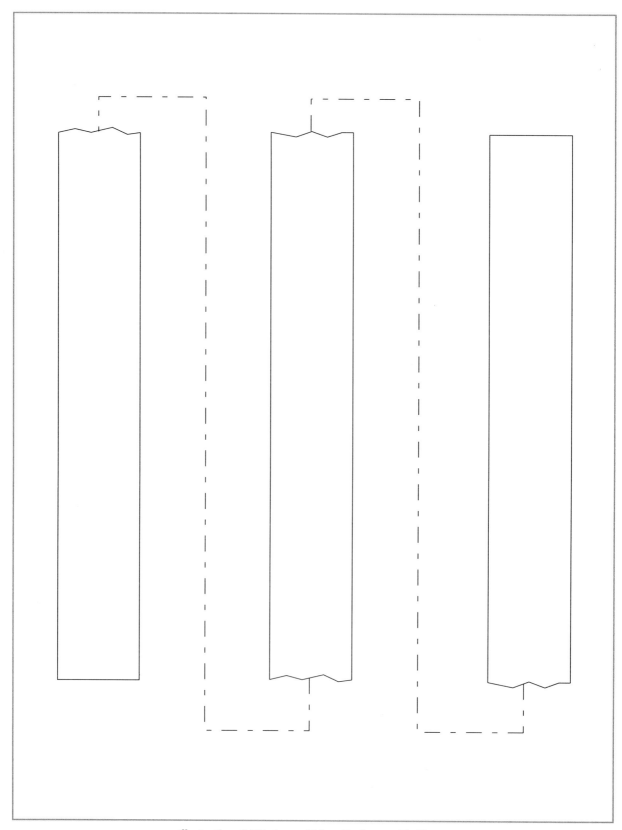

Illustration 6.10—Long Object Broken to Fit Sheet

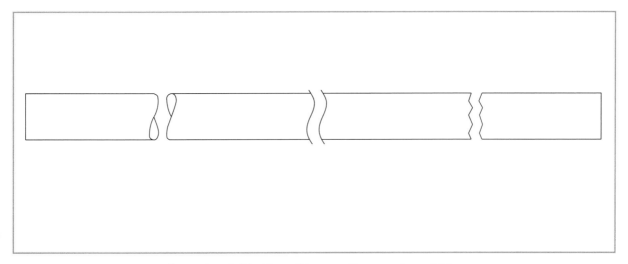

Illustration 6.11—Three Ways to Shorten a Long Object

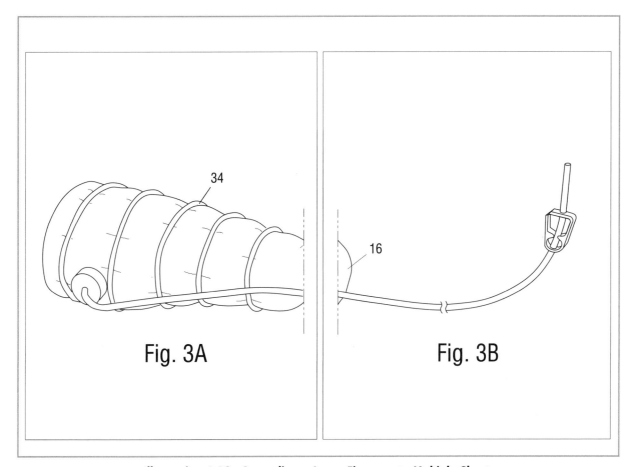

Illustration 6.12—Spreading a Large Figure onto Multiple Sheets

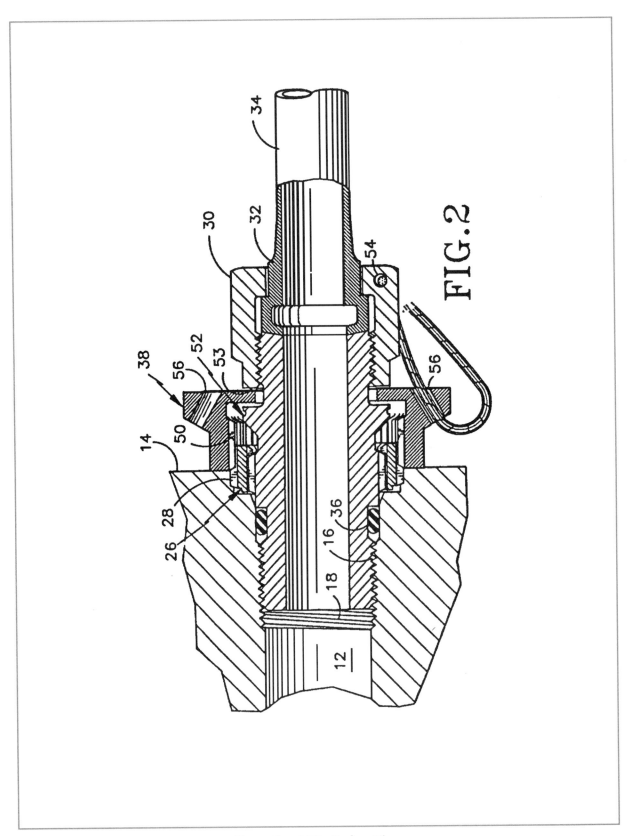

FIG.2

Illustration 6.13—Broken View

For example, the broken line in Fig. 1 indicates that the object is sliced at about its midpoint. The arrows indicate that, to see the sliced section as it appears in Fig. 2, a person must be standing to the right of the object and looking toward its left end. The number "2" is placed at the tip of each arrow, to indicate that the sectional view is Fig. 2. The broken line may be referred to in the specification (description portion of the application) as "line 2 – 2."

5. Perspective General View

Instead of an orthogonal general view, such as in Illustration 6.14, you can use a perspective general view, which is easier to understand. However, in a perspective general view, the sectioning plane may be positioned at any angle to the drawing sheet, so greater care should be taken to make its orientation clear. Illustration 6.15 shows how the broken line and arrows are drawn in a perspective general view:

1. "Slice" the object with an imaginary, rectangular sectioning plane. The dashed lines are only for illustrating the plane and where it meets the object; they should not appear in a patent drawing.
2. Draw the broken line on the sectioning plane. Imagine threading a string through the object; the broken line is the part of the string that remains visible.
3. Position the arrows at the ends of the broken line, so that they are perpendicular to the imaginary plane, and pointed in the direction of the view.
4. The finished broken line with arrows as it should appear in a patent drawing, without the imaginary plane.

The sectioning plane in the illustration shown is horizontal with respect to the object, but a sectioning plane may be arranged in any orientation desired.

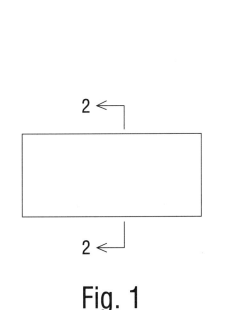

Fig. 1

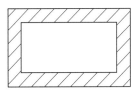

Fig. 2

Illustration 6.14—Sectional View

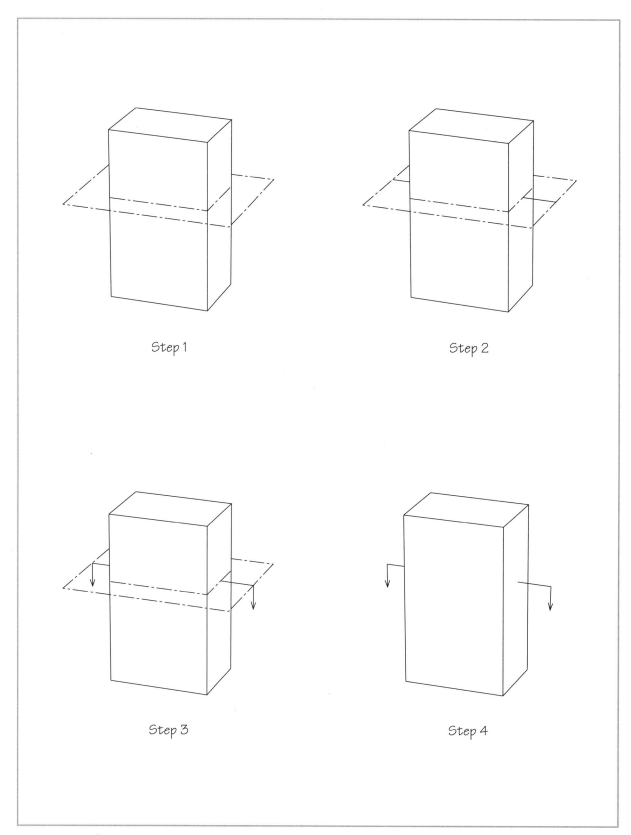

Step 1

Step 2

Step 3

Step 4

Illustration 6.15—Positioning Broken Line and Arrows in Perspective General View

6. Partial Sectional View

In contrast to Illustration 6.14, which shows the object completely sliced in half, Illustration 6.16 shows a partial sectional view, in which a motor or generator is shown with a portion cut away. The delineation between the sectional and general view is preferably jagged to clearly show that it is a sectional cut. The non-sectioned portion of the motor or generator provides a built-in general view, so that no broken line and arrows are needed. Partial sectional views may be used when it is not necessary to show all of an object's interior.

7. Orthogonal vs. Perspective Sectional

The easiest type of sectional view to make is the orthogonal sectional, such as those shown in Illustration 6.14 and Illustration 6.16, in which the viewer is looking at the exposed surface at a right angle, or head-on. However, just as with other orthogonal views, orthogonal sectional views tend to be difficult to understand, particularly if the object depicted is complicated. A much clearer type of sectional view is the perspective sectional, such as Fig. 3 in Illustration 6.17. The general view for a perspective sectional should also be a perspective of the same angle, which in this illustration is Fig. 2. A broken line with arrows is not necessary in this case, because the orientation of the sectioned object, and where the section is taken, are clear when the two views are compared.

8. Line Types and Hatching

The "exposed surfaces" of a sectioned part must be covered by hatching, which are oblique parallel lines. Crisscross hatch lines, such as a weave pattern, cannot be used. The angle of hatching is ideally 45 degrees from horizontal. The hatching in adjacent parts should be at opposite angles to clearly show that they represent different parts. If there are more than two adjacent parts, their hatch styles and angles can be varied to distinguish them, as shown in Illustration 6.18. Different areas of a continuous part, such as the two ends of the sectioned ring in Illustration 6.19, should have the same hatch style, spacing, and angle.

The hatch lines should be thinner than the edge lines (lines representing the outline and edges of the object) to avoid confusion. In Illustration 6.20, the figure with thinner hatch lines is clearly easier to understand than the one with identical line widths throughout.

9. Enlarged Views

A portion of a device may be enlarged in a separate figure to show details. Dashed circles are used to surround the enlarged portion in the general view and the enlarged view, as shown in Illustration 6.21. A reference number in the general view, which is "8" in this illustration, is applied to the dashed circle to indicate the figure number of the enlarged view, which is Fig. 8 in this illustration.

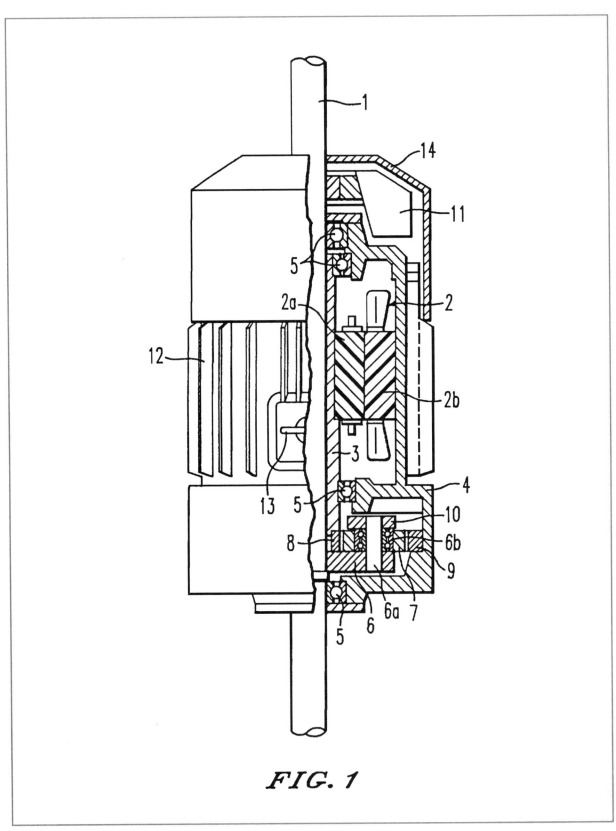

FIG. 1

Illustration 6.16—Partial Sectional View

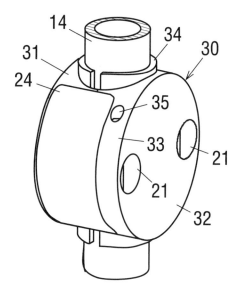

FIG. 2

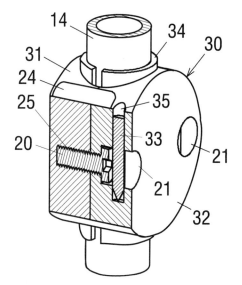

FIG. 3

Illustration 6.17—Perspective Sectional View

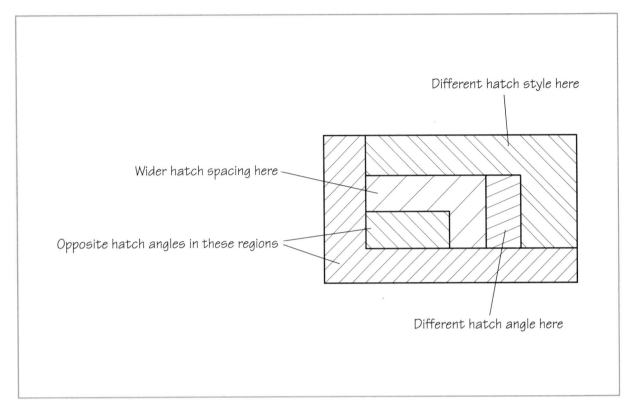

Different hatch style here

Wider hatch spacing here

Opposite hatch angles in these regions

Different hatch angle here

Illustration 6.18—Vary Hatching to Distinguish Adjacent Regions

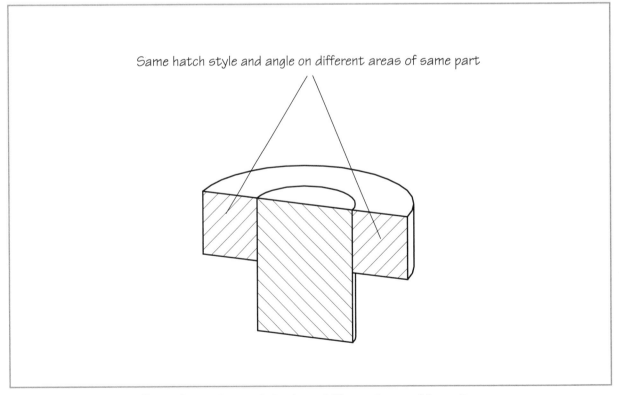

Same hatch style and angle on different areas of same part

Illustration 6.19—Hatch Angle on Different Areas of Same Part

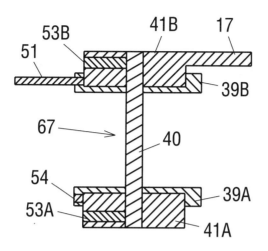

Thick Hatch Lines Make Drawing Confusing

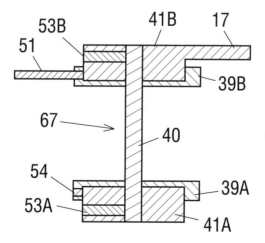

Thin Hatch Lines Make Drawing Clearer

Illustration 6.20—Thin vs. Thick Hatch Lines

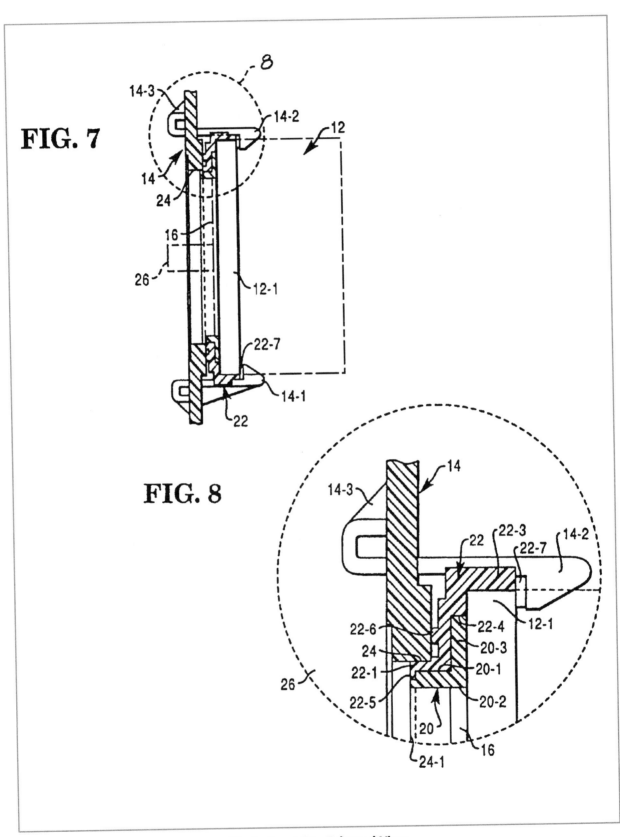

FIG. 7

FIG. 8

Illustration 6.21—Enlarged View

C. Inventions With Moving Parts

An invention with moving parts should be shown with the parts positioned in an initial or at-rest condition. Movement of the parts may be shown in several different ways.

Arrow. If the movement of a part is simple, such movement may simply be indicated by an arrow, such as the rotation of the hinged arm in Illustration 6.22.

Dashed part. If it is necessary for comprehension, a part may be drawn in solid lines to show it in its initial position, and drawn in phantom lines (dash-dot-dash lines) in the same figure to show it in a moved position. This should be done only if there is no risk of confusion. Illustration 6.23 shows both the original and moved positions of a sleeve on a rod.

Separate figures. The initial position of the part is shown in solid lines in one figure, and its moved position is shown in solid lines in a separate figure, such as the sleeve and rod in Illustration 6.24. The movement of the part may be indicated by an optional arrow.

Long sequence. A complicated invention should be illustrated with a sequence of figures to clearly show the movements and interactions of all the parts in as many distinct steps as necessary. Illustration 6.25 to Illustration 6.32 show the different figures for illustrating a fairly complicated invention—a dual-cycle toilet flusher that provides selectable small and big capacity flushes.

Fig. 1 of Illustration 6.25 is a general view that shows the device installed in a conventional water tank. Note that the toilet bowl and the water tank, which are not part of the invention, are shown in phantom (dash-dot-dot-dash) lines. Elements that

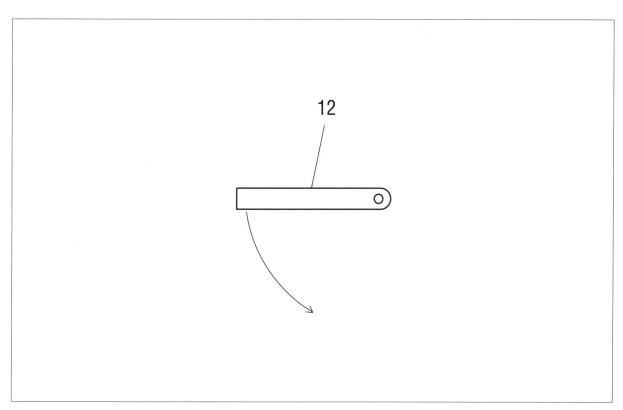

Illustration 6.22—Arrow Indicating Movement

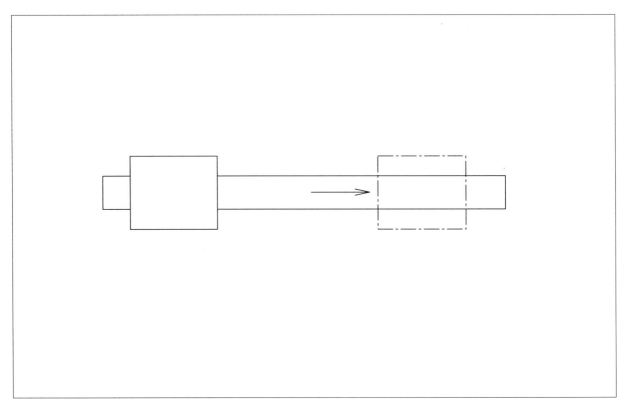

Illustration 6.23—Representing Moved Part With Phantom Lines

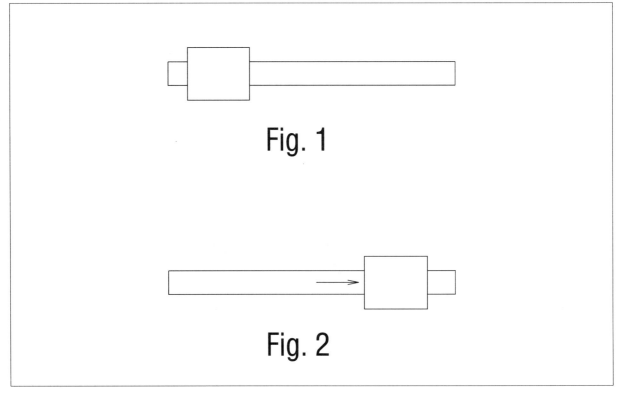

Illustration 6.24—Representing Moved Part With Separate Figures

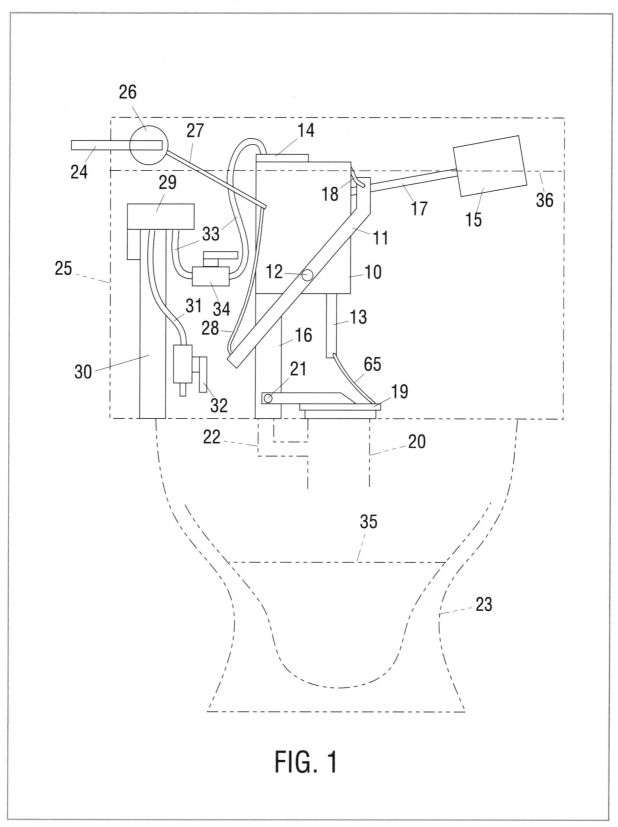

FIG. 1

Illustration 6.25—Illustration of Structure and Operation of Complicated Invention

are not part of the invention may be, but are not required to be, shown in phantom lines. Because the flushing mechanism is inside the tank, using phantom lines for the tank enables the flushing mechanism to be shown in solid lines. On the other hand, if the tank is shown in solid lines, the flushing mechanism would have to be shown in hidden (dashed) lines, which would make it harder to draw and much harder to read. Another alternative is to show the flushing mechanism in solid lines, and the tank cut away to expose the flushing mechanism. However, this would have required a separate, non-cutaway general view with a broken line and arrows to indicate the sectioning plane and view direction (see Section E, above, on sectional views). Therefore, in this case, showing the tank and toilet bowl in phantom lines saves work and makes the invention easier to understand.

Fig. 2A of Illustration 6.26 is a detailed general view of the flushing mechanism alone in an initial or at-rest condition. It is provided in addition to Fig. 1, because the flushing mechanism in Fig. 1 is too small to show all of its details. Showing the flushing mechanism alone in a separate figure allows it to be drawn big enough to occupy the entire sheet, so that all the details can be clearly seen. Note that Fig. 2A is an easy-to-understand perspective view. Also note that water level 36 is shown as a phantom line. Fig. 2B is a top sectional view of a cam assembly 67, which includes nested, rotating elements on both sides of the flushing mechanism. Showing cam assembly 67 in a top sectional view allows both sides, as well as the relationship of all the nested elements, to be clearly shown. Without the top sectional view, several additional perspectives views, which are more difficult to draw, would have to be used to convey the same information. Note the broken line with arrows in Fig. 2A that indicates the sectioning plane and view direction.

Figs. 3A and 3B of Illustration 6.27 show the flushing mechanism after the flush lever (not shown) has been pressed for a small flush—that is, after the first distinct step in the operation of the mechanism. The device is shown in Fig. 3A with some outer parts omitted to show the position of the inner parts. The description part of the application should note that specific parts are omitted for clarity. For illustration, "In Fig. 3A, housing 10, trigger 43A, cam 39A, and disc 41A (shown in Fig. 2A) are omitted to clearly show the inner mechanisms they would otherwise obscure." The back sides of cam assemblies 67 and 68 are shown, in a separate rear view in Fig. 3B, in positions that correspond to those shown in Fig. 3A. Note that it would not have been possible to clearly show the operation of the mechanism with an orthogonal side view or side sectional view, because of the many layers of parts. Note that drain hole 20 is unconnected to the rest of the device, so a bracket is used to "enclose" them and indicate that all parts belong to the same figure; the bracket is not needed if drain hole 20 touches the rest of the figure.

Illustration 6.28 to Illustration 6.32 show subsequent steps in the operation of the flushing mechanism. Additional outer parts are omitted to show the movement of deeply buried inner parts. The omitted parts can only be omitted if they have been shown in previous figures, and are not essential for the understanding of the current figure. Note that Fig. 5 is taken from the opposite side, that is, a rear view, to show the operation of the parts on that side. The written description (specification) should state clearly the side from which a view is taken, and mention that parts are omitted to show others. The sequence of figures thus cooperate to clearly show the physical structure and every distinct step in the operation of the flushing mechanism.

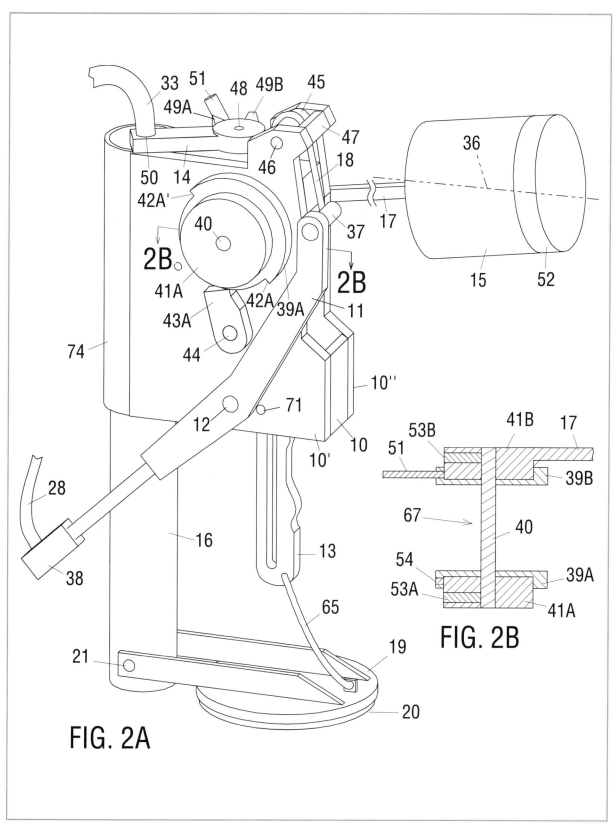

Illustration 6.26—Illustration of Structure and Operation of Complicated Invention

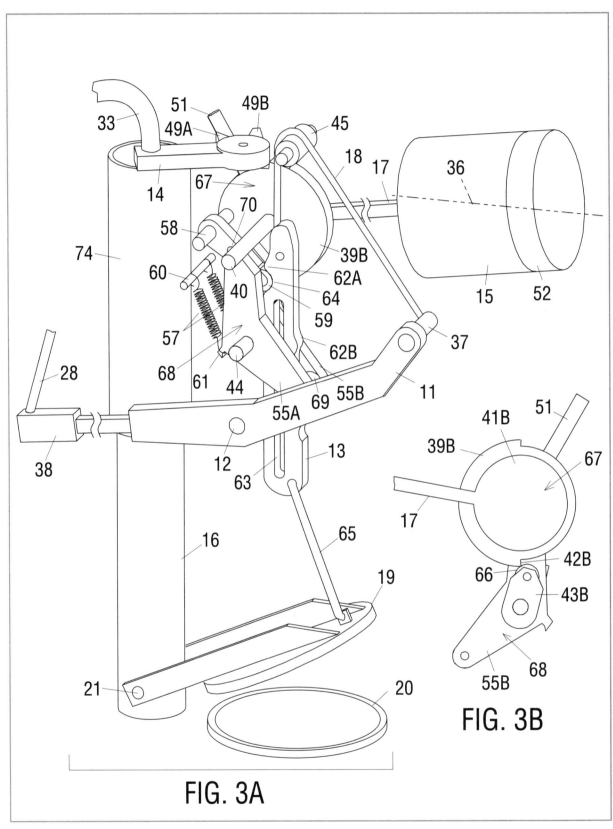

FIG. 3A

FIG. 3B

Illustration 6.27—Illustration of Structure and Operation of Complicated Invention

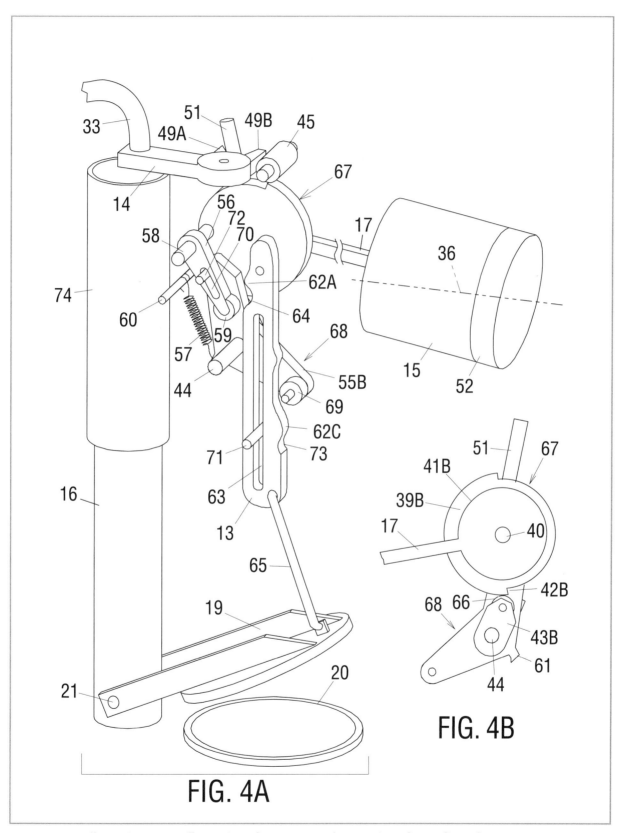

FIG. 4A

FIG. 4B

Illustration 6.28—Illustration of Structure and Operation of Complicated Invention

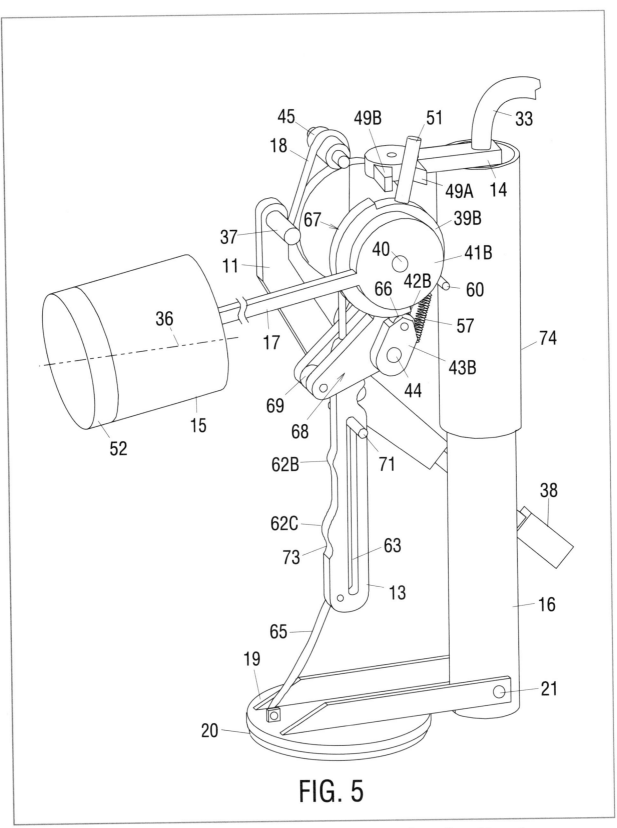

FIG. 5

Illustration 6.29—Illustration of Structure and Operation of Complicated Invention

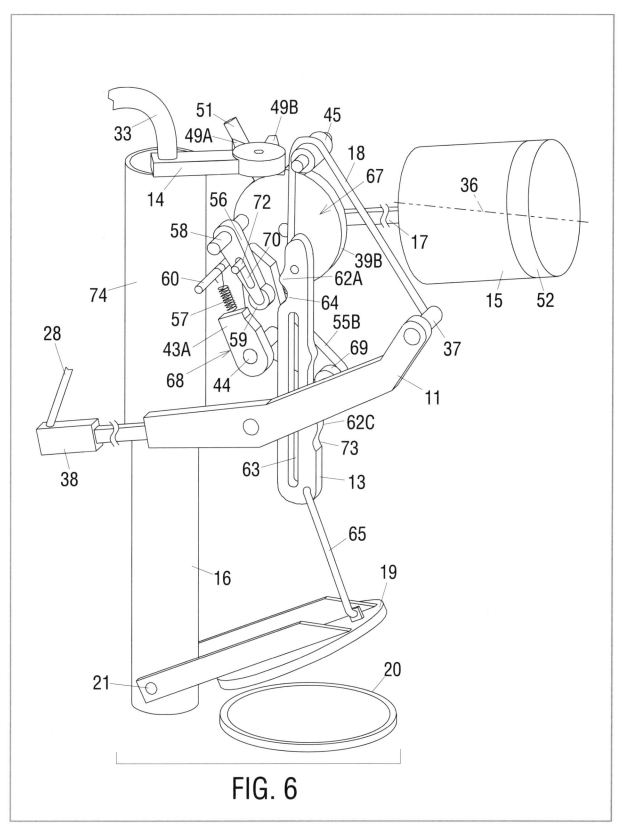

FIG. 6

Illustration 6.30—Illustration of Structure and Operation of Complicated Invention

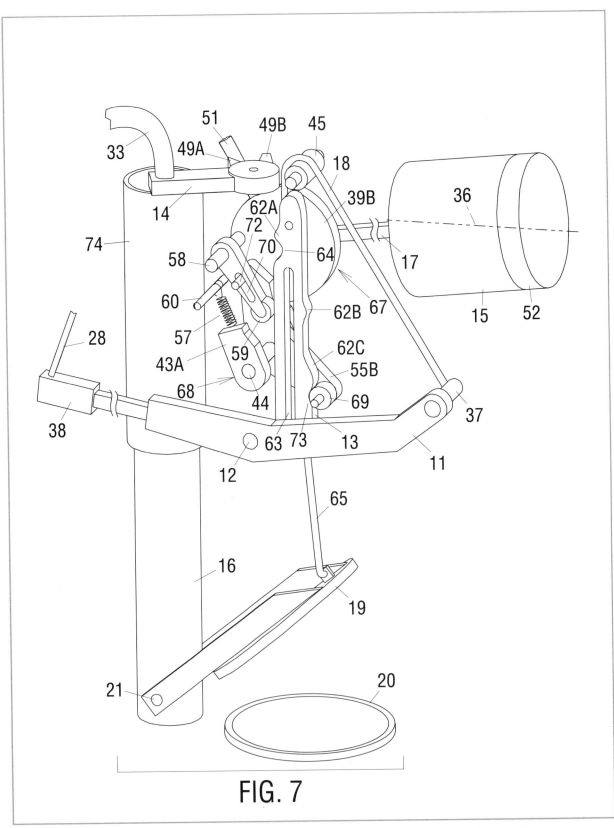

FIG. 7

Illustration 6.31—Illustration of Structure and Operation of Complicated Invention

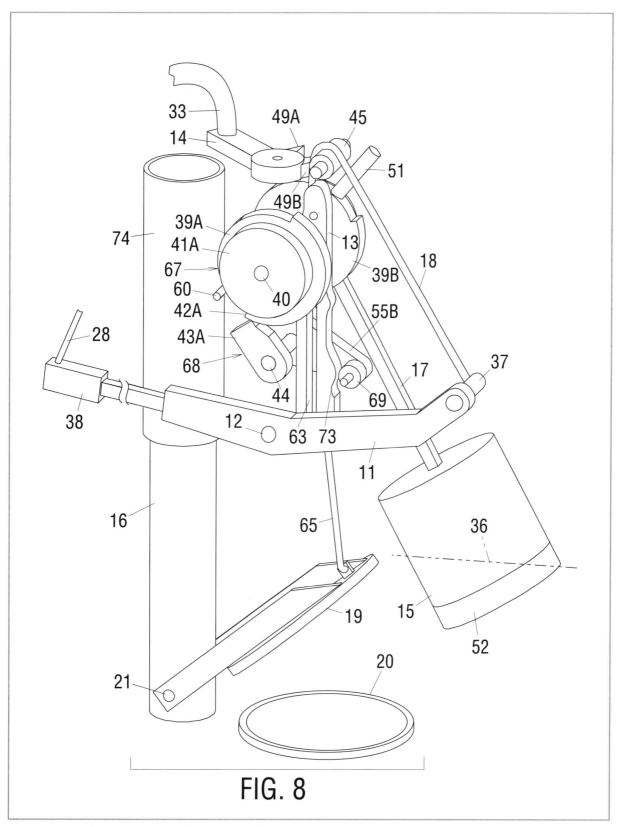

FIG. 8

Illustration 6.32—Illustration of Structure and Operation of Complicated Invention

D. Shading

Shading is the representation of shadows on an object to depict surface contour. It is distinguished from hatching, which is applied to sectioned portions of an object. (See Section C8, above.)

Shading is not required in utility patent drawings, but it can improve comprehension of the figure in some situations. Consider the unshaded cylinder in Illustration 6.33. Is the top open or closed? The shaded versions make it clear. Flat parts may also be lightly shaded, such as the closed top of the cylinder on the lower right. The object should be shaded as if it is illuminated by a light source on the upper left. Refer to Chapter 7, Section H, for details on shading techniques.

E. Graphical Symbols

There are standard symbols for many fields of technology, including electronics, fluid power (hydraulics), computer logic, plumbing, and process flow. Symbols are also considered drawings, so that they are subject to the basic drawing requirements discussed in Chapter 8, particularly Section H. There are too many types of symbols to cover here comprehensively, but some common classes are discussed below.

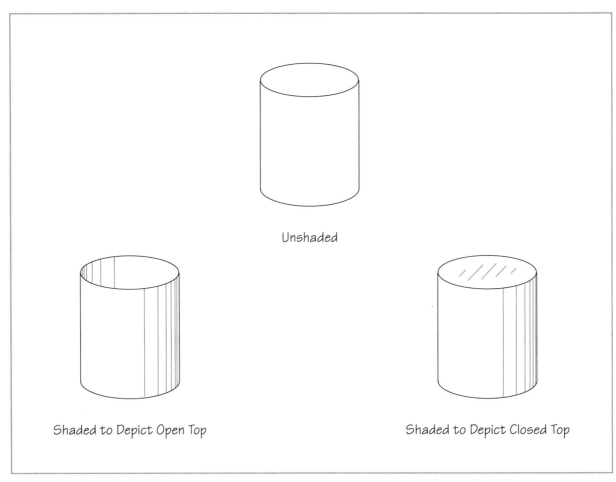

Unshaded

Shaded to Depict Open Top

Shaded to Depict Closed Top

Illustration 6.33—Shading Improves Comprehension

1. Electronic Schematics

A simple electrical circuit incorporated into a mechanical device, such as the timer shown in Illustration 6.34, may be drawn as actual parts and wires if the connections are simple enough. Alternatively, a separate schematic or electrical diagram may be used, such as the lamp controller circuit in Illustration 6.35. More complex electronic circuitry should always be illustrated with a separate schematic. Also, a figure may show a combination of actual parts and electronic symbols, such as the pill dispenser in Illustration 6.36.

Any element, including electrical components, mentioned in the description must be designated with a reference numeral using a lead line, as shown in Illustration 6.35. Customary letter-and-numeral designations for electronic parts may also be used, such as D3, R12, L3, Q23, IC5, C21, etc. ("D" means diode, "R" means resistor, "L" means inductor, "Q" means transistor, "IC" means integrated circuit, and "C" means capacitor.) If the value of a component—such as the resistance of a resistor—is important, it may be included in the description. Parts not mentioned in the description need not be designated with a reference number, but they may be labeled with a value in the drawing. A sub-circuit, such as that indicated by reference numeral 260 in Illustration 6.35, may be enclosed by a box in dashed line. If an electrical connection or line is mentioned in the description, it must also be designated with a reference numeral, for example, "speed input line 125 of F/V converter 230…" The line terminations—whether inputs or outputs—may be labeled, such as in Illustration 6.35. If a circuit is too large to fit on one sheet, it may be extended across several sheets. (See Section C3, above, on partial views, for details.) Although electronic schematics are used in these illustrations, the same rules and principles also apply to other types of schematics.

2. Block Diagrams

An electronic circuit may be represented by a block diagram, such as in Illustration 6.37, instead of a detailed schematic if the blocks represent conventional circuits. Generally, blocks diagrams are used to show very complex circuitry that would otherwise require huge schematics. The blocks must be designated with reference numerals and lead lines, and should be labeled with short names or descriptions positioned within the blocks. The lines connecting the blocks must be designated with reference numerals if they are specifically mentioned in the description, although it is not usually necessary to mention the lines. If the lines are not specifically mentioned, they do not have to be designated with reference numerals. A block diagram too large for one sheet may be spread across several sheets. (See Section C3, above, on partial views for details.)

3. Flowcharts

Processes such as manufacturing methods and computer programs are typically illustrated with flowcharts. The shapes of the boxes used in computer flowcharts should conform to standard practice: rounded box = connector; rectangles = process; diamond = decision; parallelogram = input/output; etc. A typical software flowchart is shown in Illustration 6.38, and includes boxes connected by arrows. Each box includes a short description. All the boxes should be designated with reference numerals and lead lines. Again, if a flowchart is too large for a single sheet, it may be spread across several sheets. (See Section C3, above, on partial views for details.)

needs caption

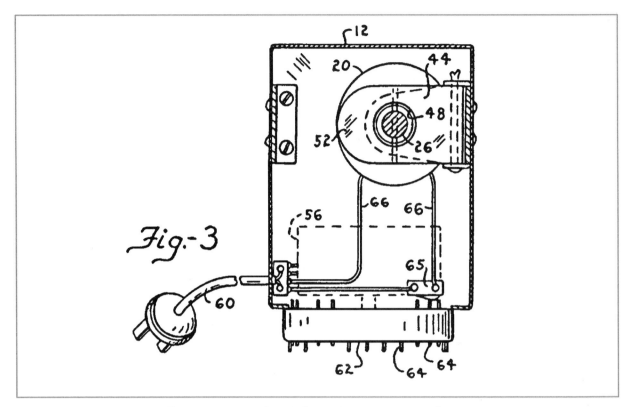

Illustration 6.34—Electrical Circuit Drawn as Actual Parts

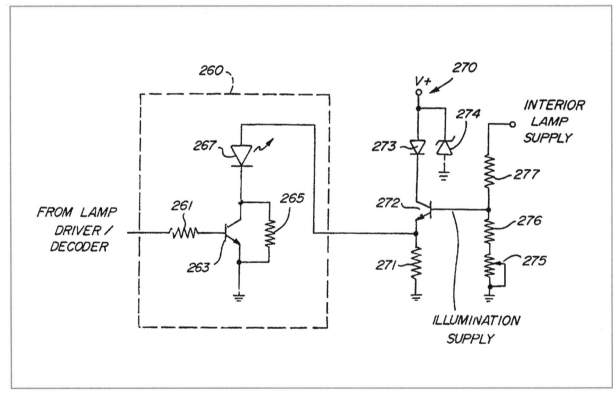

Illustration 6.35—Electronic Schematic

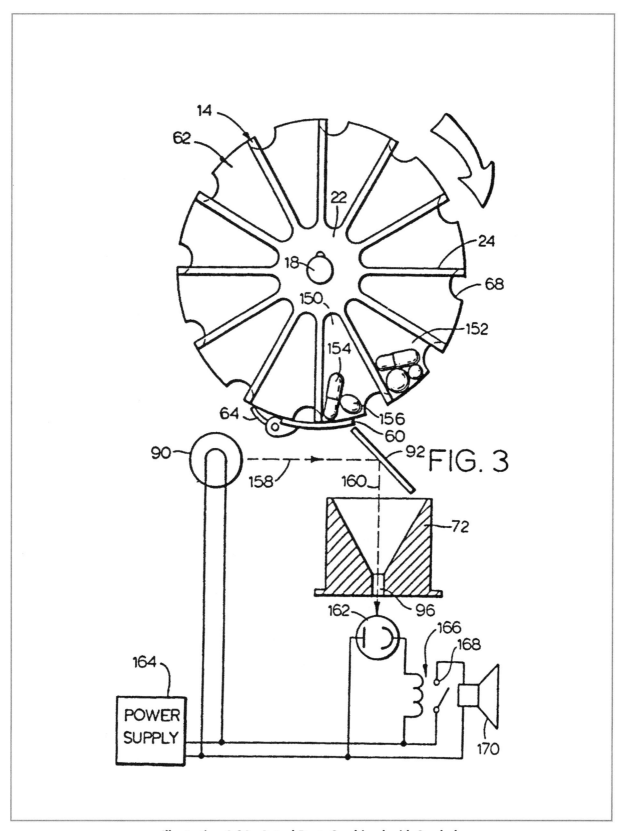

FIG. 3

Illustration 6.36—Actual Parts Combined with Symbols

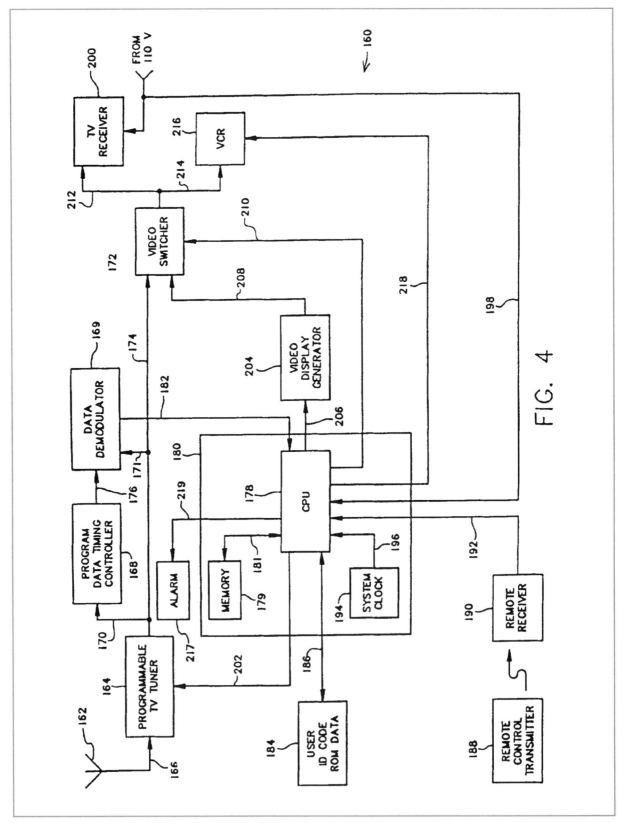

FIG. 4

Illustration 6.37—Block Diagram

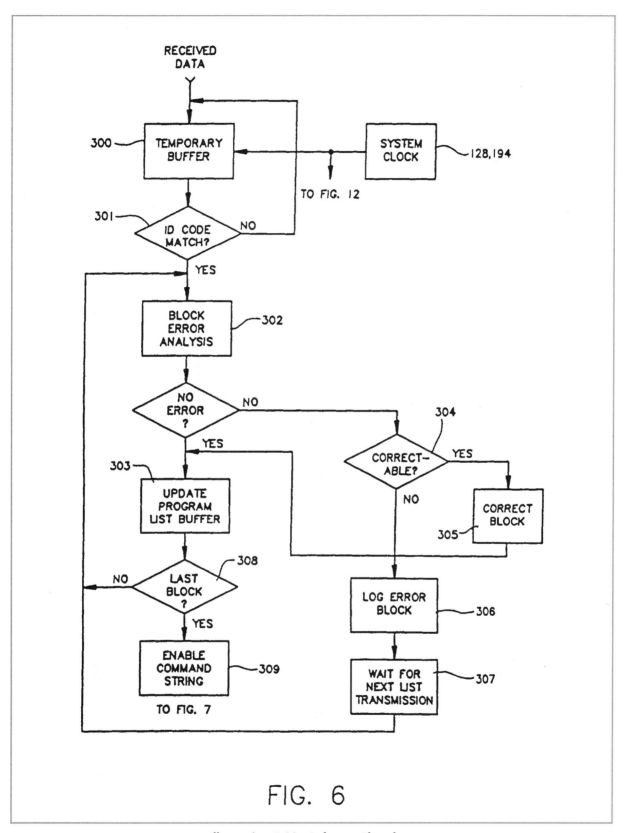

FIG. 6

Illustration 6.38—Software Flowchart

FIG. 4a

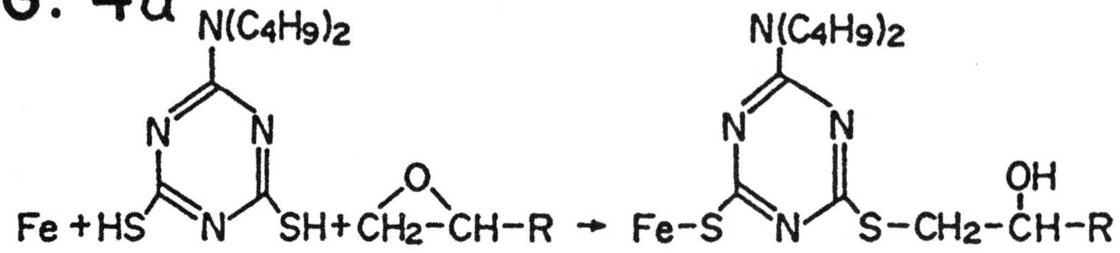

$$Fe + HS-\text{(triazine, } N(C_4H_9)_2\text{)}-SH + CH_2-CH-R \rightarrow Fe-S-\text{(triazine, } N(C_4H_9)_2\text{)}-S-CH_2-\overset{OH}{CH}-R$$

FIG. 4b

$$Fe + 3R-\text{(cyclohexane, OH, OH, OH)} \rightarrow \text{Fe complex}$$

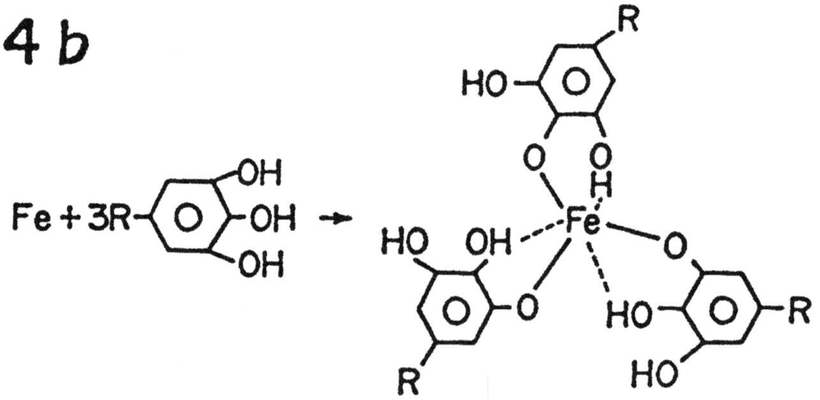

FIG. 4c

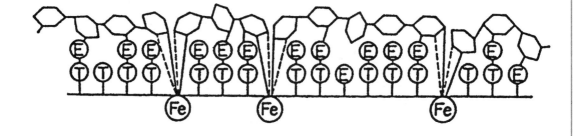

T: Triazine dithiol
E: Epoxy compound

Illustration 6.39—Chemical Formulas

4. Formulas and Tables

Chemical formulas, mathematical formulas, and tables are substantially textual information, so they may be either submitted as drawings or incorporated into the description. Only formulas may be included in the claims (legal description).

If presented as drawings, each formula and table must be labeled as a separate figure. Technically, the individual symbols and characters in a formula are not connected. However, no bracket, such as that required for exploded views, is usually required. If a formula is so complex that it may be confused as separate figures, then a bracket must be used. Typical chemical formulas are shown in Illustration 6.39, and a typical table is shown in Illustration 6.40.

If incorporated into the specification, characters used in formulas and tables must be of a block (non-script) type font or lettering style, as shown in Illustration 6.41. Capital letters must be at least 0.08" or 2.1 mm high (about 10 points), although smaller lower caps may be used. A space of at least 0.25" or 6.4 mm high should be provided between complex formulas and tables and the text of the description. Lines and columns of data in

FIG. 7

OVERALL GEAR RATIO	MAIN GEAR UNIT GEAR RATIO	AUXILIARY GEAR UNIT RATIO
FIRST	FIRST	REDUCTION
SECOND	SECOND	REDUCTION
THIRD	THIRD	REDUCTION
FOURTH	THIRD	DIRECT
FIFTH	FOURTH	DIRECT

Illustration 6.40—A Typical Table

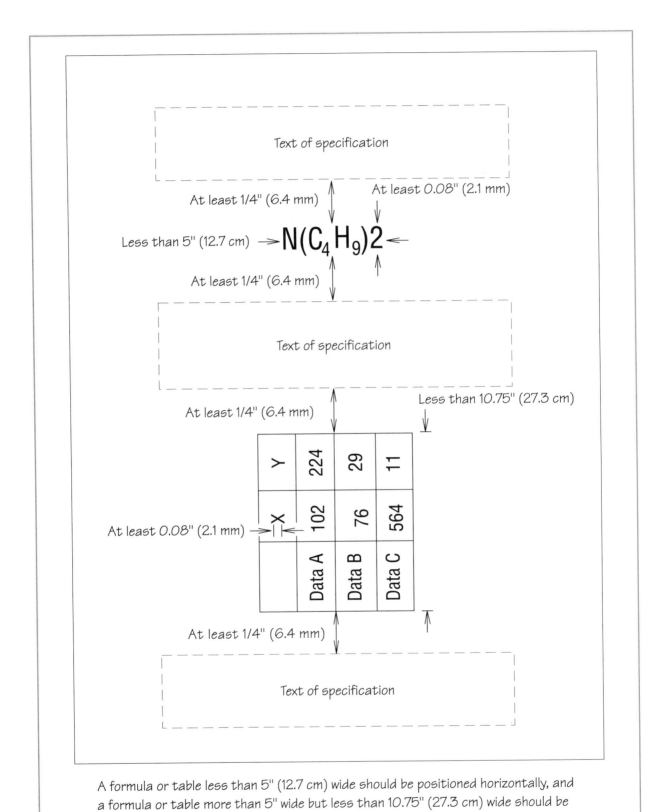

A formula or table less than 5" (12.7 cm) wide should be positioned horizontally, and a formula or table more than 5" wide but less than 10.75" (27.3 cm) wide should be positioned vertically.

Illustration 6.41—Formulas and Tables in Specification

tables should be closely positioned to conserve space. The width of a formula or table should be limited to 5" or 12.7 cm, so that it may appear within a single column in the printed patent. If it is not possible to limit the width of a formula or table to such size, it may be placed vertically, and have a maximum height of 10.75" or 27.3 cm.

5. Waveforms and Plots

Electrical waveforms and plots of other numerical relationships must be presented as drawings. A group of waveforms that illustrate events related in time must be presented as a single figure with a common vertical axis, and extend in a horizontal direction, as shown in Illustration 6.42. The horizontal direction represents time, so therefore the waveforms must be drawn to the same scale, so that signals that occur simultaneously line up properly. Alternatively, waveforms may be presented as separate figures, as shown in Illustration 6.43. Electrical waveforms, whether presented as a single or separate figures, may be connected with optional dashed lines to show relative timing relationships between them. This is the only exception to the rule that prohibits separate figures from being connected. (See Chapter 8, Section C5, for details on drawing rules.)

Each waveform must be designated with reference letters or descriptive text. The designation of each waveform may be placed anywhere next to the vertical axis, such as to the left in Illustrations 6.42 through 6.44. Different portions of interest in each waveform may be designated with reference numerals, such as in Illustration 6.44, and/or labeled with descriptive text, such as in Illustration 6.42 to Illustration 6.44.

Typical plots are shown in Illustration 6.45. Each plot must be shown as a separate figure and provided with a descriptive label.

6. Non-Standard Symbols

Standard symbols should be used whenever possible, but non-standard symbols may be used if they are not likely to be confused with standard symbols. Non-standard symbols are subject to the approval of the PTO. If you are unfamiliar with standard symbology for the field of your invention, and your invention is a very simple device, you can probably use non-standard symbols. For example, a widget with just a few electronic parts connected in a simple way can be illustrated by drawing the actual parts, or labeled rectangles to represent the parts, such as in Illustration 6.46. However, this technique cannot be used for anything but extremely simple devices, because much of the information conveyed by standard symbols cannot be conveyed by labeled rectangles.

7. Character Size

Any numeral or letter used with graphical symbols in drawings, including lower case letters in the labels, must meet the size requirement for reference characters, which is a minimum of 1/8" or 3.2 mm high (about 12 points; see Chapter 8 for details on character size requirements). It is best to use all caps, because if a mix of upper and lower case letters are used, and the lower case letters are 3.2 mm high, the caps will be too large. An exception to the size rule allows superscripts, for example, the "3" in 2^3, and subscripts, for example, the "2" in H_2O, to be smaller than 3.2 mm.

8. Descriptive Text

Descriptive text is usually not allowed in patent drawings, except in blocks such as block diagrams, flowcharts, and tables.

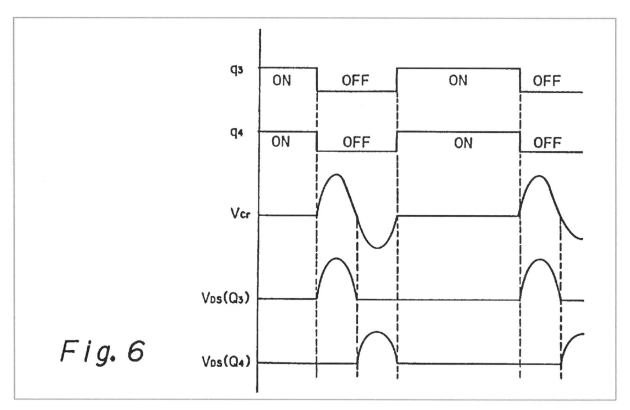

Illustration 6.42—Waveforms in Single Figure

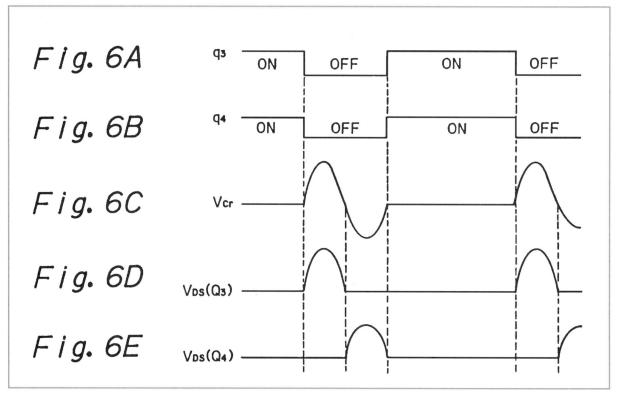

Illustration 6.43—Waveforms in Separate Figures

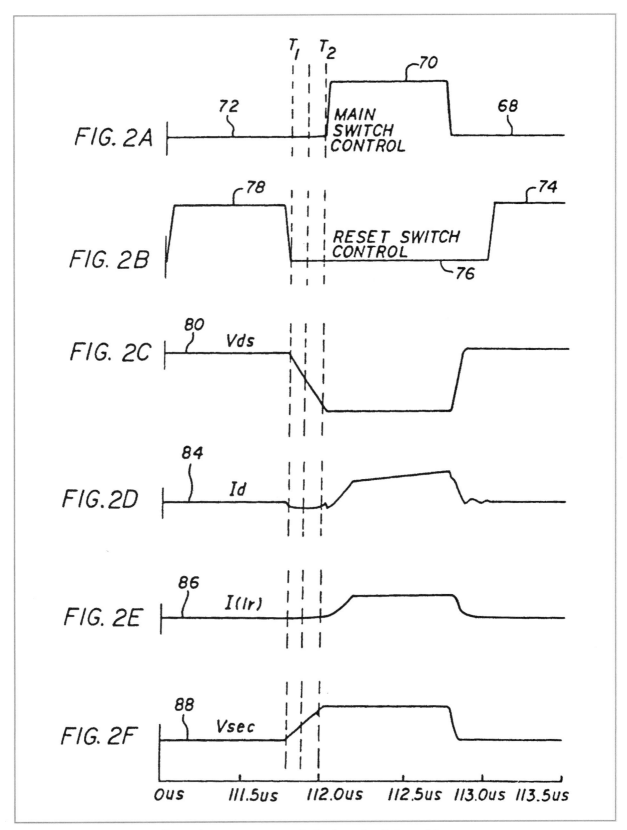

Illustration 6.44—Numerals Designating Points of Interest

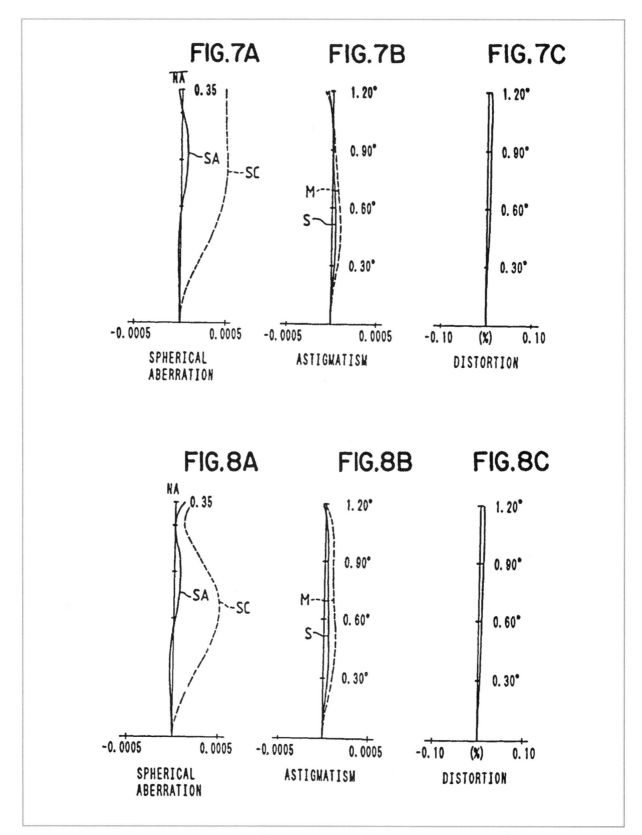

Illustration 6.45—Plots

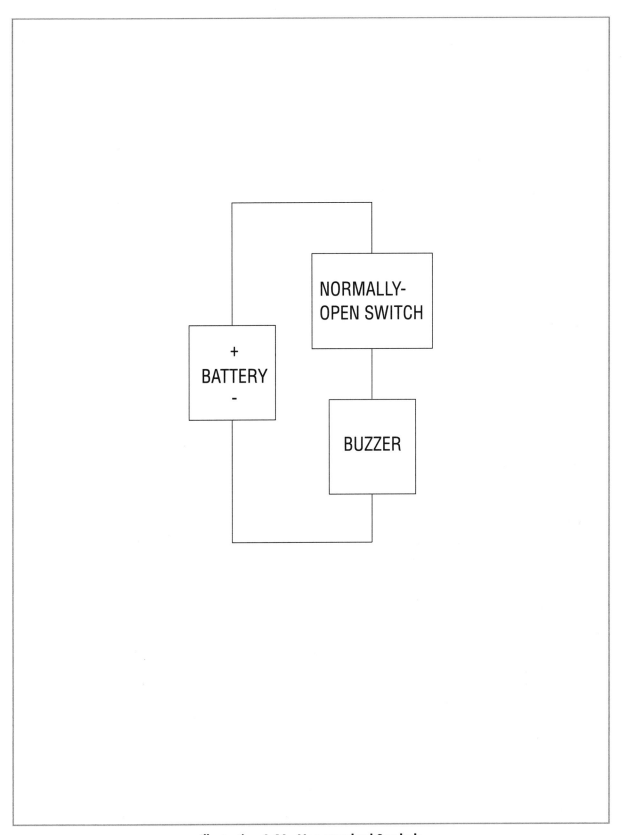

Illustration 6.46—Non-standard Symbols

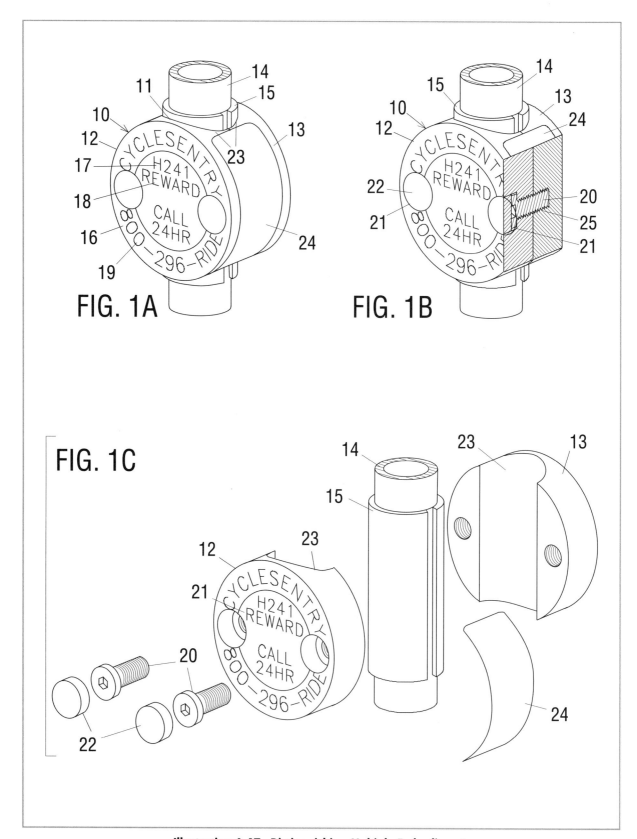

Illustration 6.47—Distinguishing Multiple Embodiments

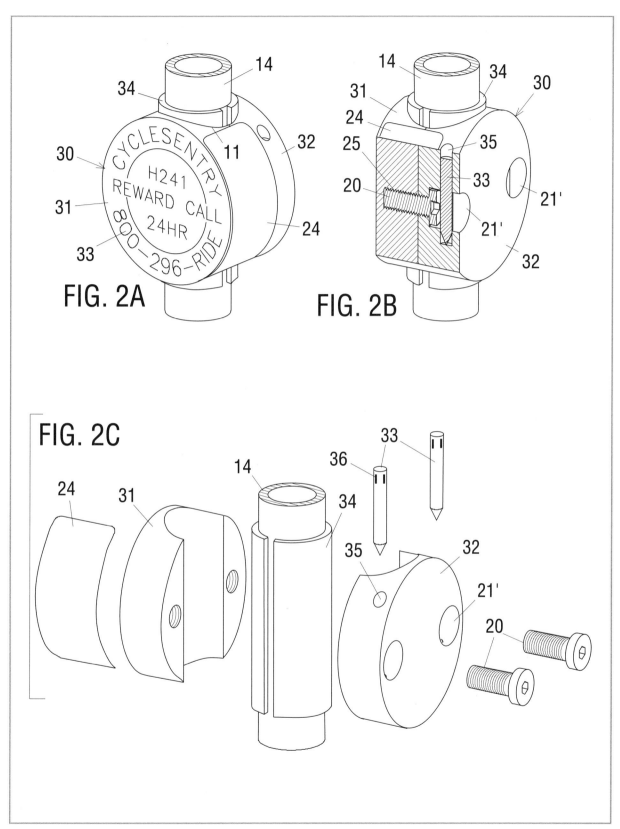

Illustration 6.48—Distinguishing Multiple Embodiments

F. Multiple Embodiments

Multiple embodiments (versions) of an invention may be included in a patent application. The different embodiments must be presented as separate figures. For example, if one embodiment of a telephone has a particular handset, and you wish to show an alternative handset, you cannot show it in the same figure as the first handset, even if you use dashed lines. The alternative handset must be shown in a separate figure, although the rest of the telephone does not have to be shown again.

We recommend using a common figure number with distinguishing letter suffixes for each embodiment. For example:

Embodiment 1:

Fig. 1A (front view)

Fig. 1B (side view)

etc. (See Illustration 6.47)

Embodiment 2:

Fig. 2A (front view)

Fig. 2B (side view)

etc. (See Illustration 6.48)

If an examiner determines that the embodiments are sufficiently different, he or she will require you to restrict the application to a single invention—that is, to choose one embodiment and cancel the figures that show the other embodiment. You may argue that the variations are not great enough to warrant treating the embodiments as separate inventions, or you may accept the requirement. If you accept the requirement, you may either file a divisional (separate) application for the canceled embodiment, or drop it. (See divisional applications in *Patent It Yourself,* Chapter 14, for details.)

G. Line Types and Width

Typical line types and widths for utility patent drawings are shown in Illustration 6.49:

1. Edge lines (lines that represent edges and corners of an object) should be continuous lines about 0.2 mm thick.

2. Hidden lines (lines that represent parts or edges that are hidden behind other parts) should be dashed lines about 0.2 mm thick. Hidden parts should be shown only when necessary so as to avoid cluttering the figure.

3. Phantom lines (lines that represent components that are not part of the invention, or a moved position of a component which is part of the invention) should be dot-dot-dash lines about 0.2 mm thick. Components that are not part of the invention may also be shown in continuous lines, which are actually preferable because they are clearer.

4. Shading lines (lines that represent shadows or surface contour) should be continuous or irregularly broken lines about 0.1 mm thick. Their thinness distinguishes them from edge lines to avoid confusion.

5. Thick edge lines (lines representing thick sheets or cords) should be about 0.5-0.8 mm thick.

6. Hatch lines (oblique parallel lines that represent sectioned parts) must be continuous lines, and should be about 0.1 mm thick. Their thinness distinguishes them from edge lines to avoid confusion.

7. Lead lines should be no more than 0.1-0.2 mm thick. A lead lines may optionally be of a type that corresponds to the part it is touching. For example, a continue lead line for a part in continuous lines, a dashed lead line for a part in dashed line, etc.

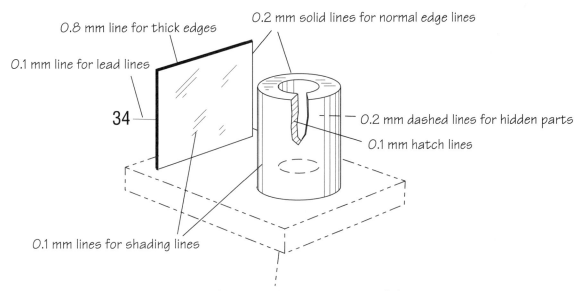

0.8 mm line for thick edges

0.2 mm solid lines for normal edge lines

0.1 mm line for lead lines

34

0.2 mm dashed lines for hidden parts

0.1 mm hatch lines

0.1 mm lines for shading lines

0.2 mm phantom lines for parts that are not part of the invention

Note that lead lines for hidden or phantom parts are drawn with corresponding line types.

Illustration 6.49—Line Types and Widths

Some Additional Points to Keep in Mind

In addition to the guidelines discussed above, here are some additional pointers:

Not too thick. Edge, hidden, and phantom lines should not be much thicker than 0.2 mm, because thicker lines tend to obscure small details. For example, if the lines are made too thick, the gap between two closely spaced parallel lines would get too small, or even disappear.

When thicker is okay. In a sectional view, edge lines may be about 0.3 mm thick, instead of the normal 0.2 mm, to more clearly distinguish them from the 0.1 mm hatch lines. Cords, cables, thick edges of sheets, or anything that is too thick to be represented by 0.2 mm lines, but not thick enough to be represented by a pair of parallel lines, may be represented by a single thick line of about 0.5-0.8 mm.

Graphical symbols. Graphical symbols—including schematics, flow charts, waveforms, etc.—should be made with continuous or dashed lines as necessary, and be about 0.2-0.3 mm thick.

■

Design Patent Drawings

This chapter details the specific requirements of formal design patent drawings. Refer to Chapter 5, Section D, for a discussion of formal and informal drawings. Also refer to Chapter 5, Section C, and *Patent It Yourself,* Chapter 1, Section B, for details on design patents, including the types of inventions that qualify for them, and how to prepare the written portion of a design patent application.

A. Amount of Detail Required

Unlike a utility patent, which covers the structure or composition of useful devices, or useful processes, a design patent covers the specific aesthetic appearance of an invention, which is solely defined by the drawings. Therefore, design patent drawings must accurately illustrate the invention's shape, proportions, surface contours, and any special material properties or textures. The drawings must show every feature of the invention that is visible during normal use, so that no part of it is left to conjecture. They must be shaded to depict surface contour or characteristics, such as transparency, and to distinguish between open and solid areas. If the drawings depict the invention inaccurately—for example, if they show incorrect proportions, contour or other details—the resulting patent may end up protecting the wrong shape. Therefore, it is extremely important that the drawings be accurate in depicting the shape you wish to protect.

Show Idealized Form

If you have made a rough mock-up of your invention that is not visually identical to the design you wish to patent, you should show the invention in its idealized form in the patent drawings. You can apply for a patent on a design even if you have not made a model. Therefore, if you have made a model, the drawings may be idealized as much as you wish, without regard to the appearance of the model. The important thing is to show the invention accurately in its ideal form.

Drawings cannot be changed after filing. *The appearance of a design invention cannot be changed after filing, not even slightly. Therefore, features cannot be added to or removed from design drawings. However, if a feature was described in the original application as not being part of the invention, or if the feature was shown in dashed lines (which are used to illustrate features that are not part of the invention; see Section G, below) in the original drawings, it may be removed if desired. See Chapter 9, Section C, for details on the rules regarding changing drawings.*

B. Views Required

A design patent application must provide a set of six standard views, including orthogonal front, back, right, left, top, and bottom views (this requirement does not apply to utility patent drawings). The entire invention must be shown in each of these views. One or more perspective views, perhaps one front and one back, should also be included if the required orthogonal views do not clearly illustrate the invention. The angle of the perspective view should be carefully selected to maximize comprehension. Note that there are no reference numbers like those typically used in utility patent drawings, because reference numbers are not allowed in design patent drawings. (See Chapter 1 for details on orthogonal and perspective views.)

As shown in the drawings of a computer mouse in Illustration 7.1, Fig. 1 is a rear perspective view, Fig. 2 is a top orthogonal view, Fig. 3 is a rear orthogonal view, Fig. 4 is a left side orthogonal view, Fig. 5 is a front orthogonal view, Fig. 6 is a right side orthogonal view, Fig. 7 is a bottom orthogonal view, Fig. 8 is a front perspective view, and Fig. 9 is a left perspective view. Although several perspective views are used in this illustration, one is usually enough for simple inventions, such as the mouse. However, inventions that are very complicated or difficult to understand should have as many perspective views as necessary.

1. Exceptions to the Rule

There are a few exceptions to the requirement for the six standard views:

1. Any view that is duplicative of another need not be shown. For example, if an invention has identical or symmetrical right and left sides, then only one side needs to be shown. The description of the figures in the specification should note such fact. For example, "Fig. 3 is a right side view of the invention; the left side view being a mirror image."

2. A view of any side of an invention which is plain and unornamented, such as the flat bottom of a lamp base, may be omitted. The description of the figures in the specification should note such fact. For example, "The invention includes a plain and unornamented bottom, which is not shown."

3. A thin and flat object, such as fabric, embossed designs, etc., would only require front and rear views. Again, the description should note such fact. For example, "The invention is a thin and flat sheet; therefore only front and rear views are shown."

2. Consistency of Views

The appearance of the invention must be consistent throughout the different views. That is, there must be no discrepancies between the views in the appearance of any element in the invention. For example, a design for a computer housing must not be shown with round button in one view and square button in another view. All details must match.

C. Drawings Must Show All Features

Design patent drawings must show every part of the invention that is visible during normal use; the only exceptions are listed in Section B, above. If an invention includes novel aesthetic features that are visible during normal use, but which cannot be illustrated by the standard views discussed in Section B, above—for example, interior surfaces— the following views may be used in addition to the standard views.

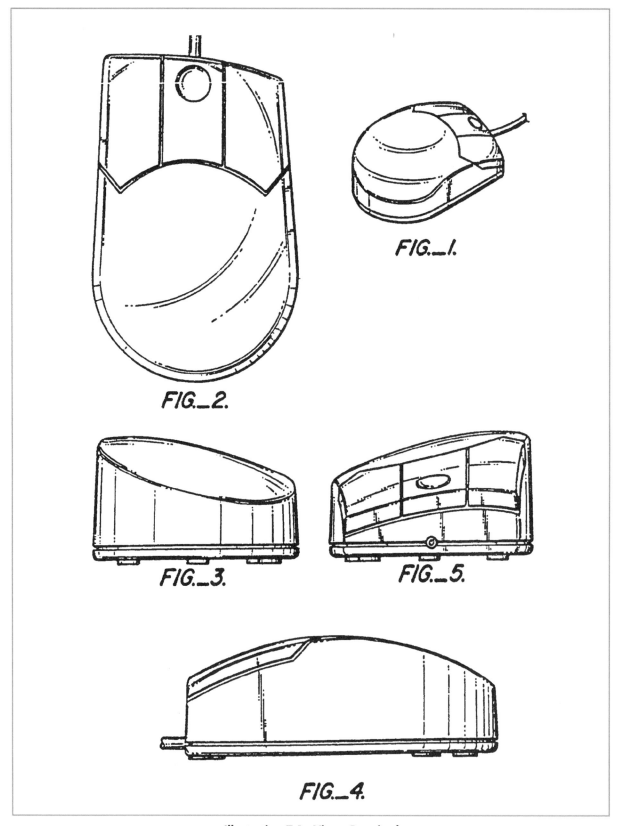

FIG.__1.

FIG.__2.

FIG.__3.

FIG.__5.

FIG.__4.

Illustration 7.1—Views Required

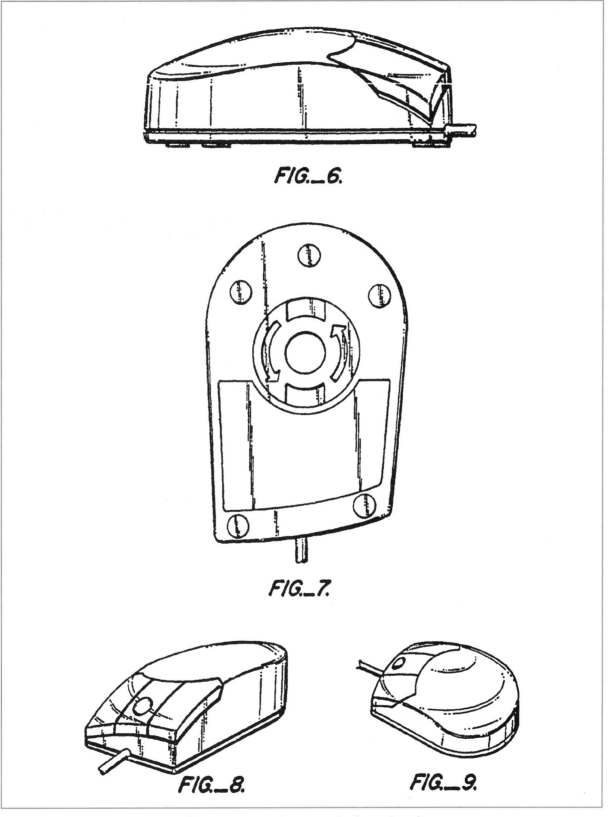

FIG.__6.

FIG.__7.

FIG.__8.

FIG.__9.

Illustration 7.1—Views Required (continued)

1. Sectional View

As shown in Illustration 7.2, in addition to the standard views (only two shown here), a glass cover is shown in Fig. 3 in a sectional view sliced in half to illustrate its ornamental internal contour. The sectional view is necessary in this case, because the internal contour of the cover would not be clear from a bottom view, which is also required, but not shown here. A broken line with arrows must be placed on a general view (a suitable non-sectional view), such as Fig. 2 in Illustration 7.2, to indicate the sectioning plane and view direction. See Chapter 6, Section C, for details on sectional views.

2. Exploded View

An invention with parts that are separable during normal use may have such parts shown separated, and independently positioned as necessary to show their internal design features, such as Fig. 2 in Illustration 7.3. In such an exploded view, a bracket must be used to "enclose" or connect the parts that belong to the same figure. Projection lines (dashed lines connecting separated parts) should not be used in design drawings. Again, in addition to the exploded view, the invention must also be shown assembled in the standard views, such as Fig. 1. (Other standard views, as discussed in Section B, above, are also required, but they are not shown in Illustration 7.3)

3. Parts Shown Separately

Normally separable parts may also be illustrated in separate figures to show their interior design features. In Illustration 7.4, the dish and lid are shown assembled together in Fig. 1, and the dish is shown alone in Fig. 2 to illustrate its contour that is normally covered by the lid. The description of

the figures should state the fact that a part is shown separately. For example, "Fig. 1 is a front perspective view of a dish with lid showing my new design. Fig. 2 is a front perspective view of the dish, shown without the lid to illustrate interior ornamental features." Again, as discussed in Section B above, the entire invention must also be shown assembled in the other normally required views (not shown here).

Caution About Purely Functional Parts

In design patent drawings, sectional or exploded views cannot be used to show purely functional internal features that are without any aesthetic value, such as working mechanisms. Of course, if an internal functional part has an ornamental shape—for example, a stylized engine of an automobile—it may properly be shown.

D. Parts Behind Transparent Surfaces

Any part that is visible behind a transparent surface should be shown in solid lines, just as it would be seen in real life. The lines representing such parts should be thinner than the other lines, to distinguish them, such as in Illustration 7.5.

E. Movable Parts

Movable parts cannot be shown in alternate positions in design patent drawings in the same figures, but they can be shown in separate figures. If an invention has normally separable parts, use sectional or exploded views as discussed above.

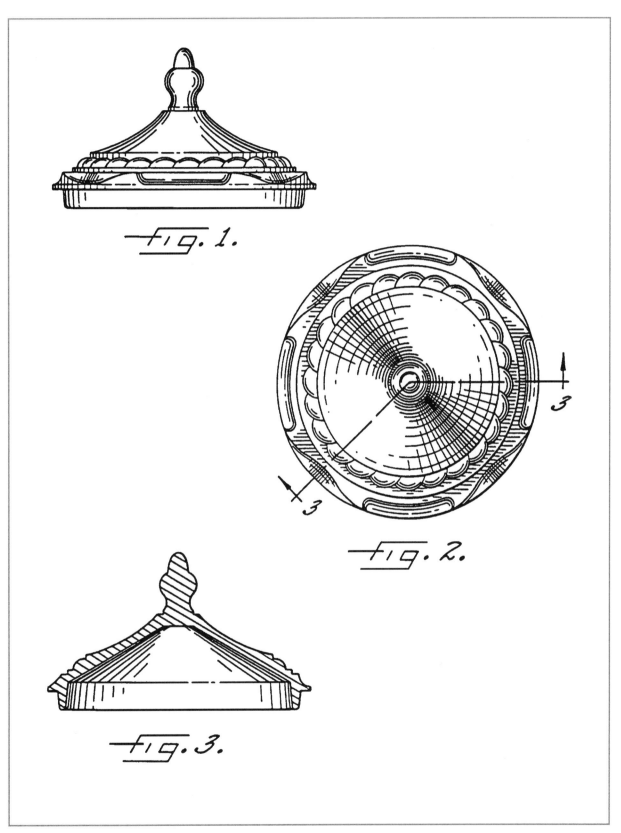

fig. 1.

fig. 2.

fig. 3.

Illustration 7.2—Sectional View

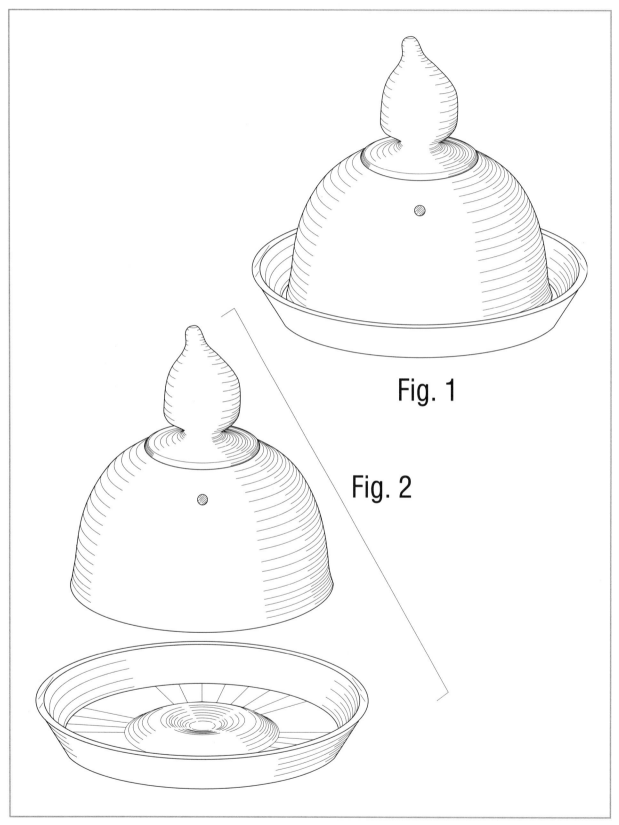

Fig. 1

Fig. 2

Illustration 7.3—Exploded View

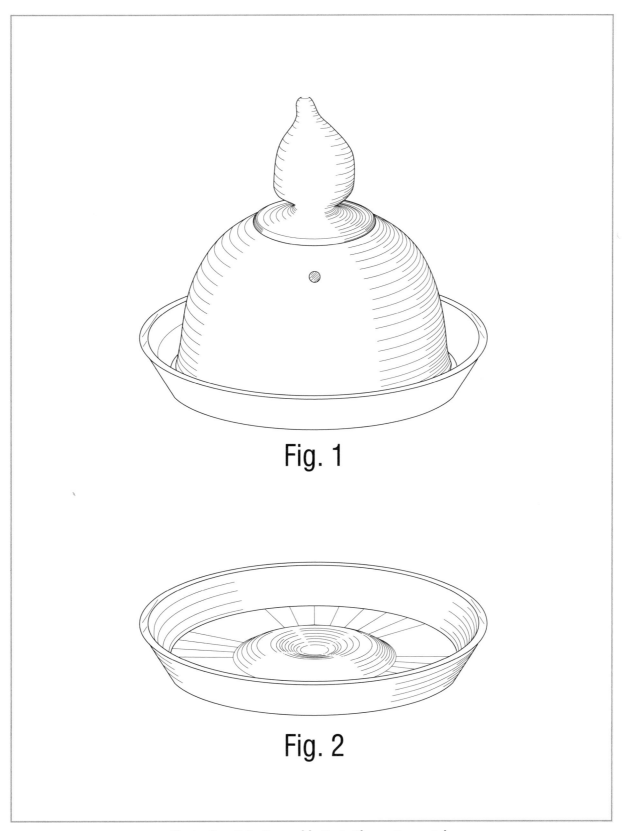

Fig. 1

Fig. 2

Illustration 7.4—Separable Parts Shown Separately

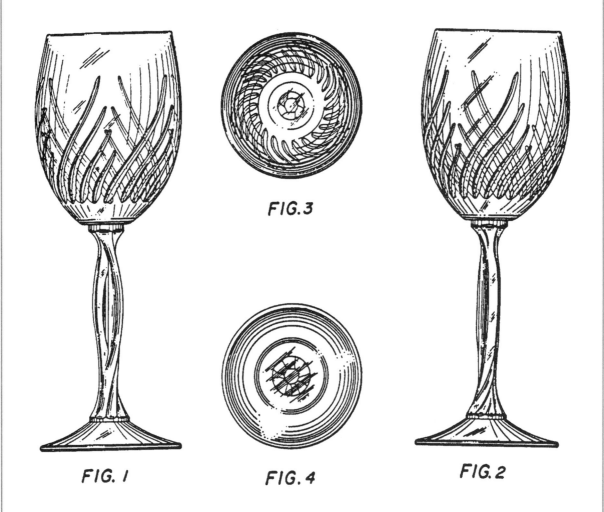

FIG. 1   FIG.3   FIG.4   FIG. 2

Illustration 7.5—Transparent Objects

F. Surface Markings

Surface markings, such as labels and logos, may be shown in continuous lines on the invention. As discussed in Section A, above, anything shown in continuous lines will be considered part of the invention. Therefore, you must consider whether such markings are an integral part of your invention before they are included. If a marking is simply one of different possible variations, such as the time displayed on a digital clock, then such marking should be shown in dashed lines. (See Chapter 8, Section G, for details on dashed lines.) If a marking is intended to be merely an example, it should be noted in the description of the figures. For example, "The marking shown in Fig. 1 is merely exemplary." You may simply omit a marking if it is not an important part of your invention.

G. Unclaimed Matter

Any element shown in continuous lines is considered as part of the invention, whereas any element shown in dashed or phantom lines (see Chapter 8, Section G, for details on dashed and phantom lines) is considered to be *not* part of the invention. Therefore, do not use dashed or phantom lines unless they represent elements that you do not wish the patent to cover (unclaimed matter)—for example, a background setting, a person using the invention, etc.

In the telephone design shown in Illustration 7.6, the base unit is shown in continuous lines and the handset is shown in dashed lines. The base unit is covered by the patent, but the handset is not. The handset is shown merely to convey the invention more clearly, but is actually unnecessary, because the base unit is clearly recognizable as such without it. Elements in dashed or phantom lines should only be shown when they are absolutely necessary to clearly convey the invention.

H. Shading Techniques

Shading is the representation of surface contour and texture, and is a very important part of design patent drawings. If a drawing is filed without shading, or with inadequate shading, the examiner will object to the drawings and require you to provide new, corrected drawings. The PTO will accept new drawings that correct minor shading problems, as long as such problems do not cause significant ambiguity in the appearance of the invention. See Illustration 6.33 in Chapter 6, Section E, for an example of ambiguity caused by the lack of shading.

If the shading, or lack thereof, causes enough ambiguity, the examiner may reject the design application as being based on inaccurate drawings. (See Chapter 9 for details on rejections by the PTO.) Such a rejection usually cannot be overcome, because it is not permissible to add any new matter (information) to an application after filing, including shading that substantially affects the shape of the invention. Therefore, you must make sure that the drawings as filed clearly depict every feature of your design, including any surface contour or texture that must be illustrated by shading. Other than on the invention, shading must not be applied to anything else on the drawing, such as the background.

1. Linear and Stippled Shading

The shading should be applied as if the light source is to the upper left of the drawing, so that the shadows are on the right and bottom sides. The two conventionally-accepted types of shading are linear and stippled. Linear shading uses parallel lines—either continuous or broken—as shown in Illustration 7.7 and Illustration 7.8, whereas stippled shading uses tiny dots, as shown in Illustration 7.9. Either type of shading is acceptable to the PTO, so choose the one you prefer. The techniques for applying shading in specific situations are discussed in the illustrations.

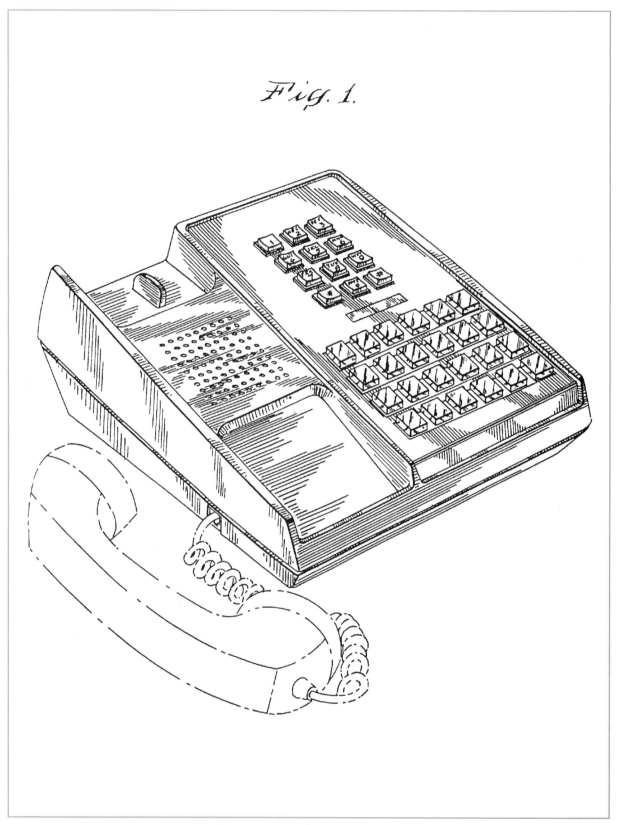

Fig. 1.

Illustration 7.6—Unclaimed Matter

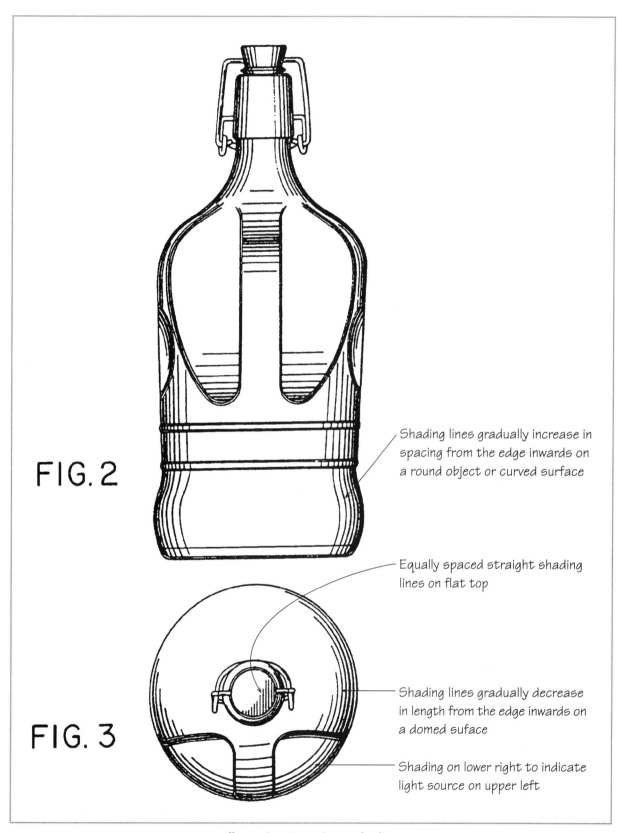

FIG. 2

FIG. 3

Shading lines gradually increase in spacing from the edge inwards on a round object or curved surface

Equally spaced straight shading lines on flat top

Shading lines gradually decrease in length from the edge inwards on a domed suface

Shading on lower right to indicate light source on upper left

Illustration 7.7—Linear Shading

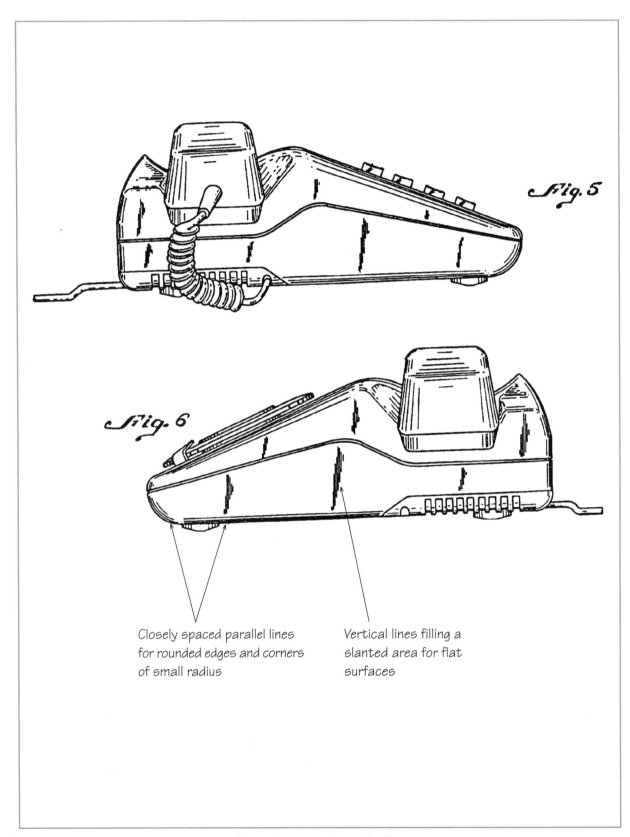

Fig. 5

Fig. 6

Closely spaced parallel lines
for rounded edges and corners
of small radius

Vertical lines filling a
slanted area for flat
surfaces

Illustration 7.8—Linear Shading

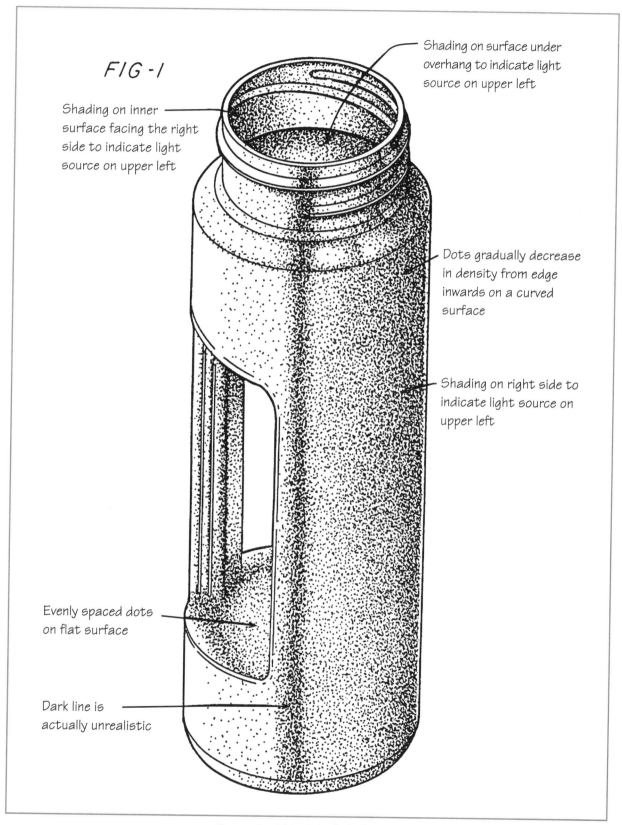

FIG-1

Shading on surface under overhang to indicate light source on upper left

Shading on inner surface facing the right side to indicate light source on upper left

Dots gradually decrease in density from edge inwards on a curved surface

Shading on right side to indicate light source on upper left

Evenly spaced dots on flat surface

Dark line is actually unrealistic

Illustration 7.9—Stippled Shading

Stipple shading is normally used for representing shadows—that is, surface contour—but it may also be used for representing rough textures, such as foam, coarse fabric, concrete, etc. Linear shading is preferably used for depicting transparent or shiny surfaces, such as glass or polished metal. Linear and stippled shading can be combined in a drawing and used wherever suitable. For example, the body of a car can be stippled shaded, and its glass can be linear shaded, such as in Illustration 7.10. Solid black shading is not allowed, except to depict a solid black color when color is an important design feature, such as on a paint scheme of a racing car. (See Section G, below, for details on the representation of color.)

2. Computer-Generated Shading

This section introduces a completely new, computer-generated shading technique for producing the most accurately-shaded patent drawings. (Refer to Chapter 3 for a basic understanding of computer-drafting, which is necessary to understand this technique.) Most 3D CAD (computer-aided drafting) programs can automatically apply shading to a 3D model, so that it appears solid, as shown in Illustration 7.11. However, a printout of such a shaded model is not acceptable as a design patent drawing, because it includes no black lines, which are required (see Chapter 8, Section B), and the continuous areas of gray, which are considered to

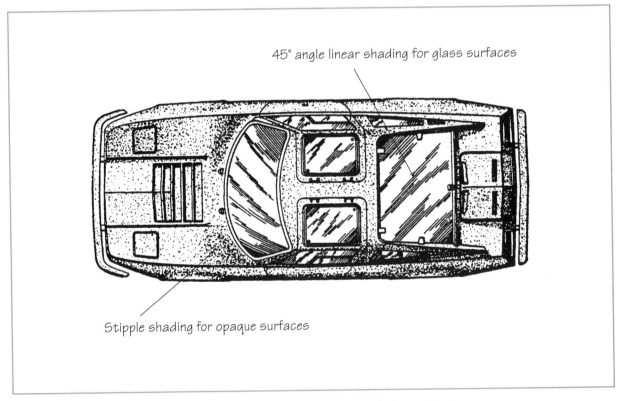

45° angle linear shading for glass surfaces

Stipple shading for opaque surfaces

Illustration 7.10—Combined Stippled and Linear Shading

be solid shading, are not allowed. The technique introduced here is used to modify such a shaded drawing, so that it may be submitted as a proper design patent drawing. Follow these seven steps:

Step 1: Build a 3D model of the invention.

Step 2: Save the 3D drawing as a 2D line drawing with hidden lines removed, and edit it as necessary.

Step 3: Apply shading to the 3D model with the Shade command in the 3D CAD program. Prior to shading, set your display to at least 1024 x 768 or higher resolution at 256 or more colors. The higher the resolution and colors, the smoother the image will appear. Lower resolutions are not suitable, because they will result in coarse images.

Step 4: Save the screen image with the Screen Save, Print Screen, or similar command.

Step 5: Merge or load the 2D line drawing on top of the shaded image.

Step 6: Scale the 2D line drawing so that it precisely matches the size of the shaded image, and line them up exactly, as in Illustration 7.12.

Step 7: Set the printer settings to 300 dpi and coarse dithering (dot pattern). An actual printout is shown in Illustration 7.13. Do not use fine or diffuse dithering, because the resulting printout will have dots that are too close together. If the printout has too many dark areas, use an image editing program, such as Corel Photopaint, to brighten (lighten) the image created in step 4, and repeat steps 5-7.

You now have a design patent drawing which is very accurately shaded.

⚠️ *As of this writing, this technique has not been used to produce a drawing submitted in an actual patent application, so we do not know for certain that the PTO will accept it. If the PTO objects to the shading, you can overcome the objection by redoing the drawing with one of the other shading techniques discussed in Section H1, above.*

I. Representation of Color and Material

Color drawings and color photographs are never accepted as formal design patent drawings, although they are accepted as informal drawings. (See Chapter 5, Section D, for a discussion of informal and formal drawings.) If the new features of your invention include the use of distinctive color schemes or materials, standard hatch patterns that represent specific colors or materials, as shown in Illustration 7.14, may be used for filling those areas. Illustration 7.15 shows a unique color scheme on a truck; the colors are represented by hatch patterns. The patterns are applied without regard to the object's surface contour, such as in Illustration 7.16, where the lines of the patterns are drawn straight even on a curved object. The patterns representing materials are applied in a similar way. This may seem to be a very inefficient way to represent colors, but it is necessary, because patents are printed only in black ink to minimize cost, and a color drawing or photograph converted into black and white by a photocopier usually becomes very difficult to understand.

When such patterns are used, the description of the drawings must state such fact. For example, "The drawings include standard drafting symbol patterns for representing color." If the particular colors indicated in the drawings are merely exemplary, the description of the figures should state such fact. For example, "The colors of the invention are not limited to those specifically indicated in the drawings." Note that these special patterns are not necessary if the colors or materials of your invention are unimportant.

Illustration 7.11—Shaded 3D Model

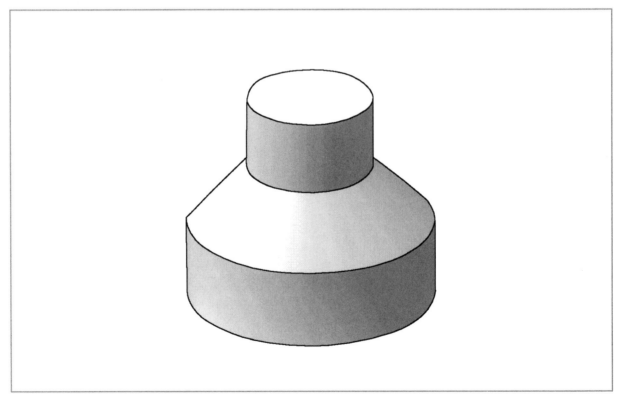

Illustration 7.12—Line Drawing Merged With Shaded Image

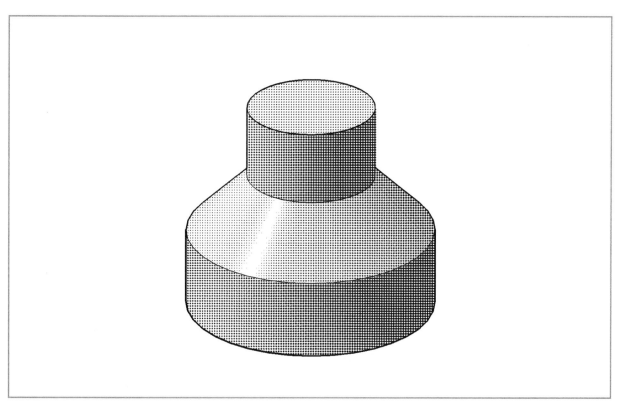

Illustration 7.13—Printout

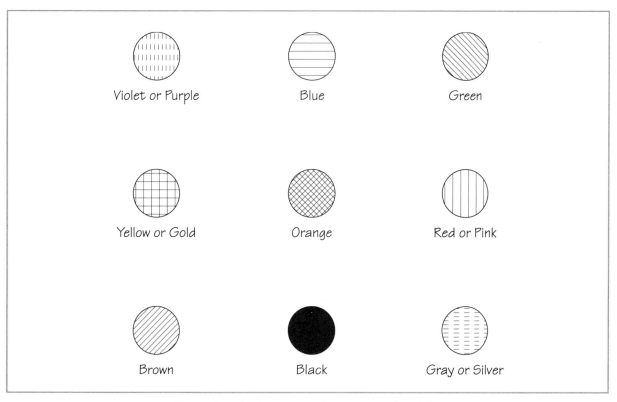

Illustration 7.14—Patterns for Representing Color

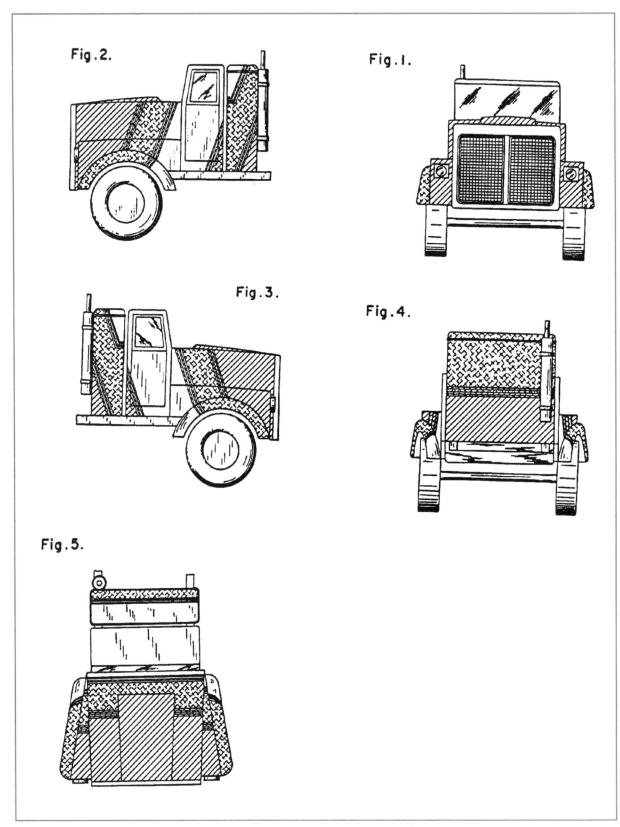

Illustration 7.15—Hatch Patterns Representing Color Scheme

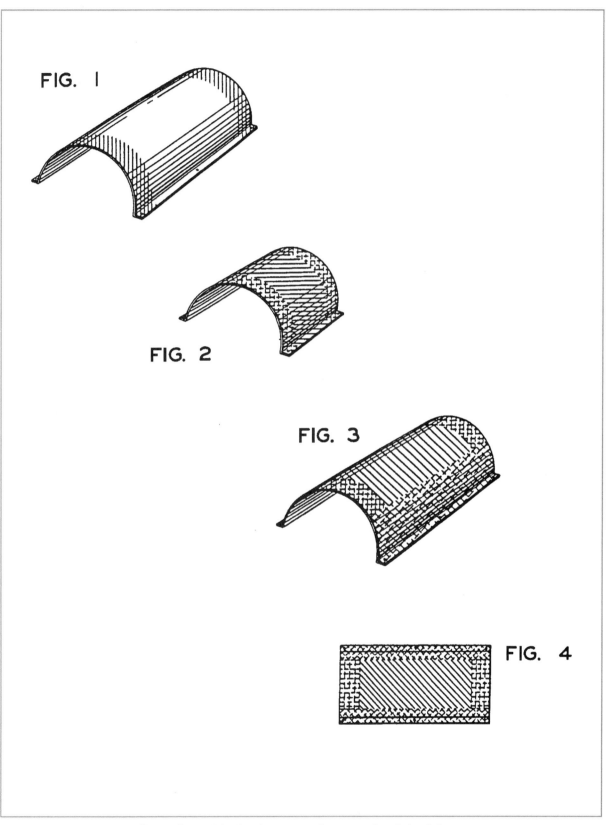

FIG. 1

FIG. 2

FIG. 3

FIG. 4

Illustration 7.16—Hatch Patterns Representing Color Scheme

J. Line Types

Design patent drawings must be made with black lines or black-and-white photographs. As already stated, color drawings and color photographs are not accepted as formal design patent drawings under any circumstances. For black line drawings, such as in Illustration 7.17, suitable line types and widths are as follows:

1. Edge lines (lines representing edges) of the invention must be continuous lines. They should be about 0.2 to 0.3 mm thick. Use 0.3 mm lines for edge lines in sectional views. Never use dashed or phantom lines for the invention.

2. Environmental structures and exemplary surface markings must be dashed or phantom lines. They should be no thicker than the lines representing the invention.

3. Shading for opaque surfaces should be parallel continuous lines, broken lines, or dots. They should be about 0.1 to 0.2 mm thick, and thinner than the edge lines. They should be oriented vertically or horizontally with respect to the object. In computer-generated shading, as described in Section F, above, the dot sizes are not directly controllable.

4. Shading for transparent or translucent surfaces should be parallel continuous or broken lines. They should be about 0.1 to 0.2 mm thick, and thinner than the edge lines. They should be oriented obliquely (slanted) with respect to the object.

5. Hatch lines in sectional views should be straight, continuous, parallel, slanted lines about 0.1 mm thick. (See Chapter 6, Section E, for details on hatching.)

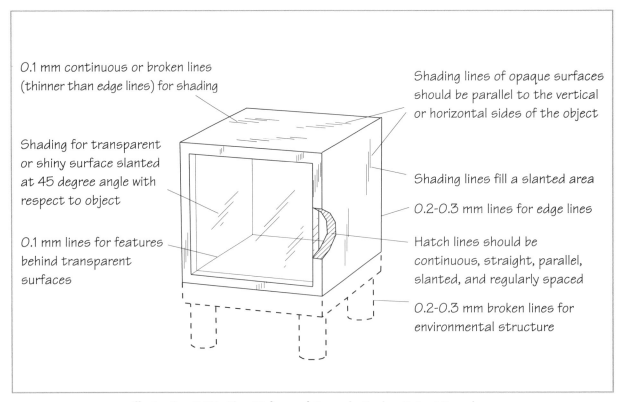

Illustration 7.17—Line Styles and Types in Design Patent Drawings

K. Photographs

Instead of black line drawings, black-and-white photographs may be submitted as informal drawings for design patent applications, and as formal drawings if the PTO grants permission (see Section K3, below). If you are adept at photography but not at drawing, it may be easier to take photographs of your invention than to make black line drawings. However, photographs must be sharp and show the invention clearly and completely, including its contours and details, without any ambiguity. Otherwise, they may be rejected for not showing the invention clearly enough. Such a rejection usually cannot be overcome without filing a new application. (See Chapter 9, Responding to Office Actions, for details on rejections.) If you are unsure about the clarity of your photographs, show them in confidence to a person who is unfamiliar with your invention, and ask him or her if all of the invention's features are visually understandable, and if there are any ambiguities.

1. Background

All photographs must have a plain background, without anything that is not part of the invention. In other words, whatever the photographs show would be considered to be part of the invention. (See Chapter 4, Drawing With a Camera, for details on taking photographs of inventions.)

2. Views Required

The views required in photographs are the same as those required in black line drawings. (See Section B, above, for details.)

3. Formal and Informal

Photographs, whether color or black-and-white, are accepted as informal drawings for design (and utility) patent applications. Black-and-white photographs are accepted as formal drawings only after the granting of a petition; color photographs or color drawings are never permitted in design patent drawings. (See Chapter 5, Section D, for details on informal and formal drawings.) Photographs submitted as informal drawings must be replaced with formal line drawings when the application is allowed (approved), or petitioned to be accepted as formal drawings. Photographs submitted as formal drawings must be printed on double weight paper or mounted on cardboard.

4. Do Not Mix Photos and Line Drawings

A design patent application may contain either photographs or line drawings, but never a mix of both as formal drawings. However, both types of drawings may be submitted together as informal drawings. Photographs submitted as formal design patent drawings must not show anything other than the invention itself.

5. Size and Margins

The size and margin requirements for photographs are the same as those for line drawings. (See Chapter 8, Section A, for details on size and margins.) Therefore, if a photograph is mounted on cardboard, the cardboard must be one of the acceptable sizes—that is, 8.5" x 11" (letter size) or 21 cm x 29.7 cm (A4 size). The photograph must be small enough to leave an acceptable border between itself and the paper, but large enough to show the invention clearly. If a photograph is submitted without the cardboard, the photograph itself must be one of the standard aforementioned sheet sizes.

6. Figure Numbers

Photographs must be labeled with consecutive figure numbers, such as Fig. 1, Fig. 2, etc. Photographs not mounted on cardboard must have the figure numbers applied directly on them. Photographs mounted on cardboard may have the figure numbers applied on the cardboard next to the photographs. (See Chapter 8, Section D, for details on numbering figures.)

L. Multiple Embodiments

Multiple embodiments (versions) of an invention may be shown in the same design patent application. Separate figures must be used for each view that shows a different design. For example, the first embodiment of a desk is shown in top perspective, front, back, top, bottom, left, and right side views. The second embodiment of the desk consists entirely of a unique inlaid pattern on the top, otherwise it is the same in other respects. The second embodiment can thus be shown with just two additional views—a top perspective view and a top view—because the other views would merely be duplicative. The description of the figures showing the second embodiment should state, for example, "Fig. 8 is a top perspective view of a second embodiment of the desk showing an alternative design of the top. Fig. 9 is a top view of the second embodiment of the desk; the other views are duplicative of the views of the first embodiment."

If the embodiments are more than just minor variations, the examiner will likely require you to restrict the application to a single invention—that is, to choose one embodiment, and cancel the figures that show the other embodiment. You may argue that the variations are not great enough, or you may accept the requirement. If you accept the requirement, you may either file a divisional (separate) application for the canceled embodiment, or drop it. (See divisional applications in *Patent It Yourself,* Chapter 14, for details.) ■

General Standards

This chapter discusses the general PTO standards for formal patent drawings for regular patent applications. Provisional patent applications, which are not examined by the PTO, can be submitted with informal drawings. See *Patent It Yourself* for details on provisional and regular patent applications, and Chapter 5, Section D, of this book for details on formal and informal drawings.

Although the drawings in existing patents are good examples of what proper drawings should look like, they cannot be entirely relied upon, because the drawing rules change from time to time, and improper drawings are sometimes allowed and printed due to oversights of the PTO. Therefore, inconsistencies between drawings in existing patents and the illustrations and rules in this book may be due to rule changes, or the drawings in question being improperly allowed.

A. Paper, Margins, and Sheet Numbering

Paper used for black line drawings must be flexible, strong, white, smooth, non-shiny, and durable. Only one side of the paper may be used.

1. Paper for Laser Output

For laser printer output from a CAD program (see Chapter 3, Drawing With a Computer, for details on CAD), use a good quality, white, smooth-surfaced laser printer paper of at least 20 lb., but preferably 24 lb. ("lb." or pound is a measure of paper thickness; an equivalent measure is called "sub" or substance. The higher the number, the thicker the paper.) A smooth-surfaced paper will provide the sharpest lines. Such paper is available at most stationary stores or large computer retailers. Avoid "computer paper"—the type with holes along the sides for dot matrix printers—and the lowest quality copier papers, which are rough and thin.

2. Paper for Ink Jet Output

For ink jet output, use a white paper specifically made for ink jet printers. Such paper prevents the ink from feathering (seeping between the fibers of the paper) so as to produce sharper images. Most laser printer paper is unsuitable for ink jet printers, because it causes the ink to feather. Ink jet papers are available at most stationary stores or large computer retailers.

3. Paper for Ink Drawings

For pen and ink drawings, use mylar film or vellum, which provide a smooth finish that prevents feathering, and are tough enough to withstand repeated erasing without damage. Such paper is generally available only at art supply stores. (See Chapter 2, Section A, for the names of two mail-order art supply vendors.)

⚠ **Beware of translucent paper.** *Ink drawings made on vellum or mylar film must be photocopied onto white paper, because vellum and mylar film are translucent (semi-transparent), and do not meet PTO standards, which require white paper. Use a copier in good condition to ensure that the copies will be clean and free of specks.*

4. Paper Size and Margins

All drawing sheets in an application must be the same size. Each sheet of paper must include an imaginary margin, free of any marks. Acceptable paper sizes and margins are shown in Illustration 8.1. The margin cannot be framed by a rectangle. The dashed rectangles in the illustrations are there only for showing the imaginary margins; they must not appear in actual drawings. The usable portion of a sheet within the margins is known as the "sight." Crosshairs, as shown in Illustration 8.2, should be placed at opposite corners of the sight to indicate its boundaries. The centers of the

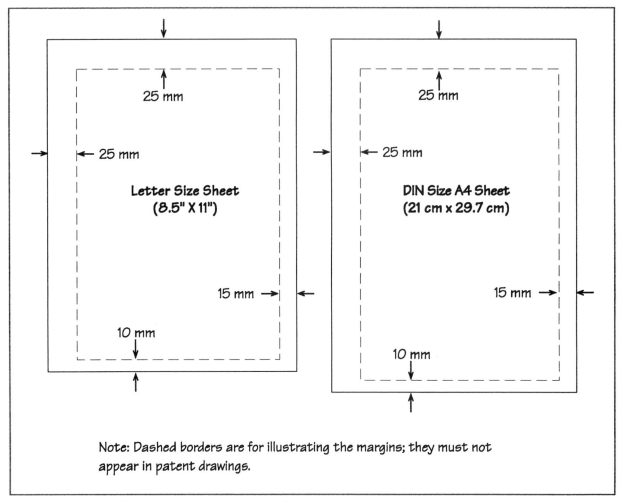

Note: Dashed borders are for illustrating the margins; they must not appear in patent drawings.

Illustration 8.1—Paper Size and Minimum Margins, Portrait Orientation

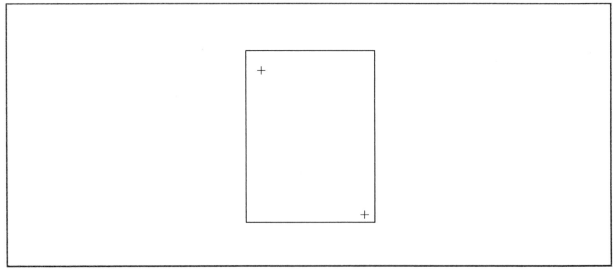

Illustration 8.2—Cross-Hairs

crosshairs should be centered on the corners, so that the outer arms of the crosshairs intrude into the margins.

Make sure the figures are positioned at least a few millimeters away from the margins, because the PTO sometimes complains of margin intrusions even when the figures are clearly inside them. Keeping the figures well clear of the margins will avoid such a problem. (See Chapter 9, Section H6, for additional details.)

5. Paper Orientation

A sheet of paper positioned so that it is taller than it is wide, such as those shown in Illustration 8.1, is said to be in "portrait" orientation. This is the preferred position. A sheet of paper positioned so that it is wider than it is tall, such as those shown in Illustration 8.3, is said to be in "landscape" orientation. Landscape orientation should be used only for long horizontal figures that cannot fit onto a sheet in portrait orientation without appearing too small.

Avoid landscape orientation whenever possible. *Do not use landscape orientation— that is, paper that is wider than it is tall—unless it is absolutely necessary. This is because, at the PTO, the drawings are attached to the same folder that also contains the written description (specification) of the invention, which is printed in portrait orientation. Therefore, drawings in landscape orientation force the examiner to turn his or her head, or rotate the entire folder to view them, which can be irritating. And of course, you don't want to irritate the examiner if you can avoid it!*

6. Sheet Numbering

The sheets of drawing paper must be numbered in consecutive Arabic numerals (the symbols 1, 2, 3, 4, etc.), starting with 1. The required format is "sheet number/total number of sheets." For example, if there are a total of four sheets, they should be numbered 1/4, 2/4, 3/4, and 4/4. No other marking—such as inventor name, title, etc.—may accompany the sheet numbers. As shown in Illustration 8.4, the sheet numbers must be placed on the top center of each sheet, below the imaginary top margin—that is, within the sight. If a drawing figure on a sheet must be positioned very close to the top center of the sheet, where the sheet number would normally be, the sheet number may be placed at the upper right corner of the sight. The sheet numbers must be larger in size than the reference numbers (discussed below in Section D) used to identify the parts of the figure—for example, about 5 mm, 1/5", or 22 points in height.

As shown in Illustration 8.5, making page numbers, figure numbers, and reference numbers different sizes allows them to be more easily distinguished.

7. Clean Paper

The paper must not be creased or wrinkled, and must be reasonably free of dirty spots, erased lines that remain visible, and other alterations. If you submit photocopies, make sure that they are substantially free of the tiny copier marks or flecks that appear on many photocopies. If a few of these marks are present, you can use white correction fluid to cover them, or make new copies after cleaning the copier's glass surface very thoroughly with soap and water or a glass cleaner.

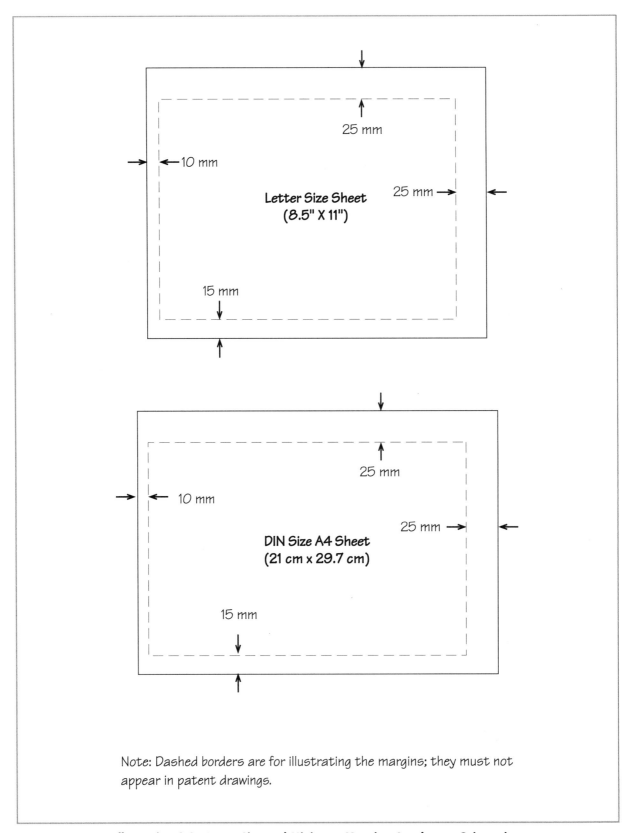

Note: Dashed borders are for illustrating the margins; they must not appear in patent drawings.

Illustration 8.3—Paper Size and Minimum Margins, Landscape Orientation

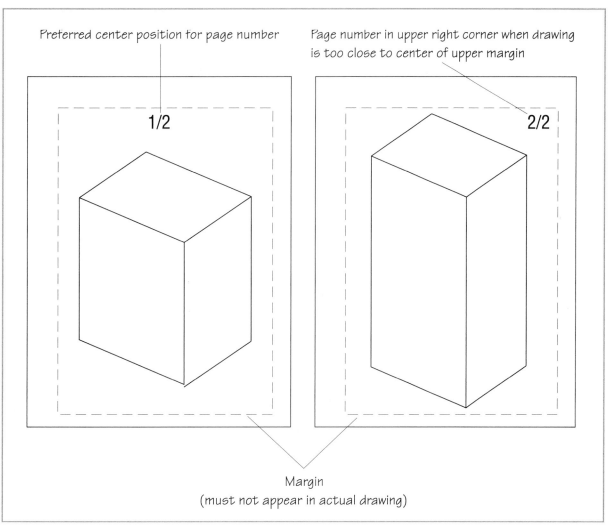

Illustration 8.4—Page Number Positioning

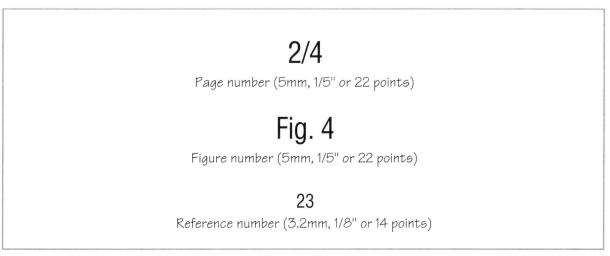

Illustration 8.5—Different Number Sizes

B. Mediums

Formal utility patent drawings may be black line drawings, black-and-white photographs, color drawings, or color photographs; formal design patent drawings may be black line drawings or black-and-white photographs. Each type of drawing and the type of inventions they are suitable for are discussed below. (See Chapter 5, Section D, for additional information on formal and informal drawings.) Plant patents are extremely rare, so they are not discussed here.

1. Black Line Drawings

Black line drawings on white paper are normally required for utility and design patent applications; they are suitable for illustrating the vast majority of inventions.

2. Black-and-White Photographs

Black-and-white photographs are normally accepted as informal drawings for utility and design patent applications, provided they show all the important details of the invention. They are accepted as formal drawings for utility and design patent applications only after the granting of a petition. The requirements for submitting black-and-white photographs as formal drawings are as follows:

1. Three identical sets of photographs developed on double-weight photographic paper, or permanently mounted on a smooth cardboard, and of sufficient clarity for reproduction;

2. A Petition for Submitting Black-and-White Photographs (a tear-out Petition is in the Appendix and a sample is shown in Illustration 8.6); and

3. A petition fee ($130 as of this writing).

No reason needs to be stated in the petition for submitting photographs. The petition will be granted automatically, provided the fee is paid and three sets of acceptable photographs are submitted. Photographs must meet the same size requirements set forth in Section A, above. For example, the cardboard on which photographs are mounted must be one of the standard sizes listed in Section A. If the photographs are submitted without cardboard, the photographic prints must be one of the standard patent drawing sizes.

Formerly, black-and-white photographs were permitted only to illustrate inventions that could not be shown by black line drawings, such as crystalline structures, metallurgical microstructures, textile fabrics, grain structures, and ornamental effects. However, due to a recent rule change, black-and-white photographs may be used for any invention, provided a petition is submitted and a fee is paid. If you contemplate using black-and-white photos instead of black line drawings, weigh the petition fee, photography cost, and the potential difficulty of making proper photos against that of making black line drawings.

3. Color Photographs or Color Drawings

Color photographs or color drawings in utility and design patent applications are normally accepted as informal drawings. Submitting color photographs instead of black-and-white ones is more convenient, because most photo developers only provide color prints. In rare cases, an invention cannot be adequately illustrated with black line drawings or even black-and-white photographs—for example, the computer-generated images that, when seen at a glance, appear to be closely-spaced dots of random colored patterns, but when viewed with unfocused eyes form three-dimensional scenes. These images must be shown in color because their hidden scenes are very subtle, and are virtually impossible to show as black lines drawings or black-and-white photographs.

App. No: _____

Filing Date: _____

Applicant: _John Smith_____

App. Title: _Table lamp_____

Examiner: _____

Art Unit: _____

Petition for Submitting Black-and-White Photographs

Assistant Commissioner for Patents
Washington, DC 20231

Sirs:

Applicant hereby respectfully petitions that the black-and-white photographs [X] filed herewith ☐ already filed be accepted as formal drawings. The petition fee is filed herewith.

Sole/First Applicant: _John Smith_____

_John Smith_____   _10-11-96_____
Sole/First Applicant Signature   Date

Joint/Second Applicant: _____

_____   _____
Joint/Second Applicant Signature   Date

Sole/Third Applicant: _____

_____   _____
Sole/Third Applicant Signature   Date

Sole/Fourth Applicant: _____

_____   _____
Sole/Fourth Applicant Signature   Date

Illustration 8.6—Sample Petition for Submitting Black-and-White Photos

Color photographs or color drawings are accepted as formal drawings in utility patent applications—never design applications—only after the granting of a petition. Unlike petitions for black-and-white photographs, which are always granted, petitions for color photographs or color drawings are granted only if a very good reason is provided to explain why color is necessary. The requirements for submitting color drawings or color photographs as formal drawings are as follows:

1. Three sets of color drawings or color photographs; color photographs must be developed on double-weight photographic paper, or be permanently mounted on smooth cardboard, and be of sufficient clarity for reproduction;

2. A petition (see Illustration 8.7 for a sample petition) that includes an explanation of why the color drawings or color photographs are necessary. (Use the tear-out Petition for Submitting Color Photographs or Drawings in the Appendix.);

3. A petition fee ($130 as of this writing); and

4. The Brief Description of the Drawings section in the specification (written description of the invention) must contain the following passage as its first paragraph:

    "The file of this patent contains at least one drawing executed in color. Copies of this patent with color drawing(s) will be provided by the Patent and Trademark Office upon request and payment of the necessary fee."

If the petition requesting to submit color drawings or color photographs is filed after the filing date of the application, so that the original specification does not include the above paragraph, a proposed amendment must accompany the petition to insert the paragraph. (See *Patent It Yourself,* Chapter 13, for the procedure for amending an application.) The petition will be granted only if the PTO determines that a color drawing or color photograph is the only viable medium to adequately illustrate your invention.

If your petition for submitting a color drawing or photograph is granted, your patent will be printed with a black-and-white copy. The color drawing will be provided only upon request and payment of a fee, as stated in item four of the requirement above.

If your petition for submitting color drawings or color photographs is denied, the examiner will object to the drawings as being improper and require you to either cancel the color photograph or drawing, or to provide substitute black line drawings. However, canceling the photo or drawing may not be an option, because the remaining drawings, if any, may not be adequate to allow comprehension of your invention. If this is the case, you must provide substitute black line drawings instead. See Chapter 3, Section 3, for details on converting photographs into black line drawings.

EXAMPLE: An application was filed with three color drawings. A petition for submitting the color drawings was denied. The drawings cannot be canceled, because the invention cannot be understood with the written description alone. In this situation, the color drawings must be converted into black line drawings.

C. Arrangement and Numbering of Figures

The drawings must be arranged and numbered in particular ways, as discussed below.

1. Arrangement of Figures

The drawings must be on sheets that are separate from the written description in the patent specification. The only exceptions are formulas and tables, which may either be submitted as drawings on separate sheets, or be incorporated into the written description as textual information. See Chapter 6, Section F, for details on formulas and tables.

App. No: _____

Filing Date: _____

Applicant: _John Smith_____

App. Title: _Field of dots with hidden image_____

Examiner: _____

Art Unit: _____

Petition for Submitting Color Photographs or Drawings

Assistant Commissioner for Patents
Washington, DC 20231

Sirs:

Applicant hereby respectfully petitions that the color photographs [X] filed herewith ☐ already filed
be accepted as formal drawings. The petition fee is filed herewith.

These color photographs or drawings are necessary because _the hidden image cannot be seen if the_
 drawing is rendered in black lines on a white background, or as a black-and-white photo.

_____

_____

_____

_____

_____

_____

Sole/First Applicant: _John Smith_____

___John Smith_____   ___10-11-96_____
Sole/First Applicant Signature   Date

Joint/Second Applicant: _____

_____   _____
Joint/Second Applicant Signature   Date

Sole/Third Applicant: _____

_____   _____
Sole/Third Applicant Signature   Date

Sole/Fourth Applicant: _____

_____   _____
Sole/Fourth Applicant Signature   Date

Illustration 8.7—Sample Petition for Submitting Color Photographs or Drawings

Each sheet of paper may contain several drawing figures, also known as views, as shown in Illustration 8.8. Each figure should be shown upright with respect to the top of the paper, depending on whether the paper is in portrait or landscape orientation. The figures must also be far enough apart to be clearly separated, so as to avoid confusion. Separate figures must not be connected by construction lines (dashed lines showing how the parts fit together). An exception to this rule is made for electrical waveforms, which may be connected by dashed lines to show relative timing, as shown in Chapter 6, Section F5, Illustration 6.43.

Bigger is clearer. *Make full use of all the available space on the paper to make the figures as large as possible. If an object must be drawn so large to show its details that there is no room for another figure, use additional sheets for the other figures. Never crowd the figures.*

2. Numbering of Figures

All figures must be numbered in consecutive Arabic numerals (the symbols 1, 2, 3, 4, etc.), starting with 1. The figures should be arranged on the sheets so that the lowest numbered views appear on sheet 1, and progress through the sheets as the figure numbers rise. The first figure number of a sheet should continue from the last figure number of a previous sheet, and must not restart from 1 on each sheet. For example, the first sheet may contain figures 1, 2, and 3, the second sheet may contain figures 4 and 5, the third sheet may contain figures 6, 7, and 8, etc. The figures on each sheet are preferably, but not necessarily, numbered so that they progress from top to bottom, as shown in Illustration 8.8.

The figure numbers must be preceded by "FIG." or "Fig."—for example, "Fig. 3." Figure numbers may include letter suffixes, such as "Fig. 1A," "Fig.

1B," etc. However, it is good practice to reserve letter suffixes only for partial views (see Chapter 6, Section D) or alternative embodiments (see Chapter 6, Section G). Each figure number must only refer to a single figure. For example, it is improper to number a sheet "Fig. 3," and number each figure "Fig. 3A," "Fig. 3B," "Fig. 3C," etc., because "Fig. 3" does not refer to a specific figure. In such a case, just leave out "Fig. 3."

3. Numbering of Partial Figures

If a figure is so large and complex that its details are too small if it is made to fit on one sheet, it can be spread across several sheets, as shown in Illustration 8.9. The figure must be arranged on each sheet so that the sheets can be tiled next to each other to assemble the complete figure. A dot-dot-dash line should be provided to denote the broken edge of each partial figure. Any arrangement of the sheets can be used, for example, side-to-side, top-to-bottom, and rectangular array, as long as the sheets can be assembled without ambiguity. Each partial figure should be labeled with a figure number having a letter suffix, for example, Fig. 2A and Fig. 2B. (See also Chapter 6, Section D, on partial views.)

4. Number Size and Style

A simple, easy-to-read lettering style or font should be used, as shown in Illustration 8.10. The ornate type of lettering typically used in very old patents should be avoided. The figure numbers must be larger in size than the reference numbers (discussed in Section E, below) to distinguish them. For example, the figure numbers should be about 5 mm, 1/5" or 22 points in height. Page number and reference number sizes are discussed in Sections A and D, respectively, of this chapter.

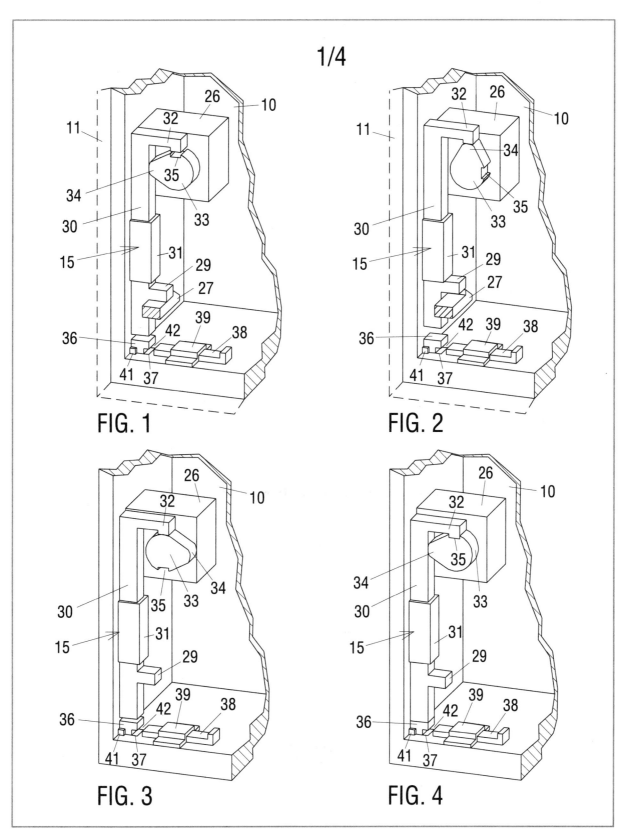

FIG. 1

FIG. 2

FIG. 3

FIG. 4

Illustration 8.8—Arrangement of Figures

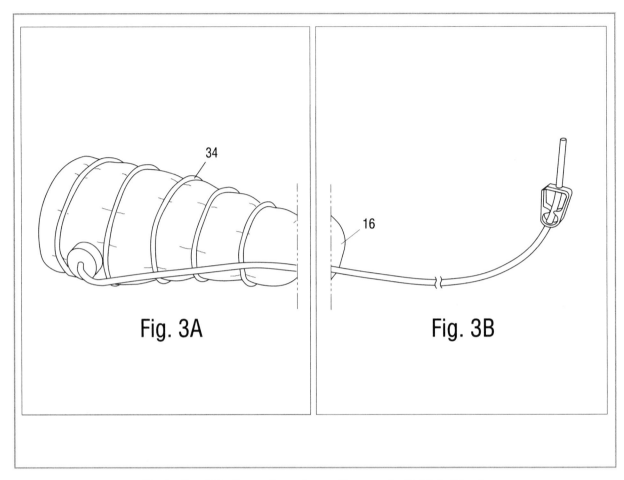

Illustration 8.9—Spreading a Large Figure onto Multiple Sheets

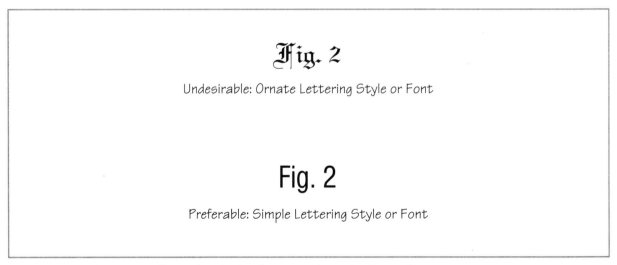

Illustration 8.10—Figure Number Lettering Style or Font

5. When a Bracket Is Necessary

If a figure includes any element that is disconnected from the rest of the figure, such as in the exploded view of Illustration 8.11, a bracket must be used to "embrace" the disconnected element to indicate that it is part of the same figure.

Even if a figure is not an exploded view, but it includes a part that is disconnected from the rest of the figure, such as part 20 in Illustration 8.12, a bracket must be used. However, if an exploded view—or any figure with disconnected parts—is the only figure on a sheet, no bracket is needed, because every element would clearly belong to the same figure.

6. Application With Single Figure

If an application includes only one figure, do not give it a figure number. The sheet number (see Section A, above) is still necessary, though, and should be "1/1."

7. Figure Depicting Prior Art

The written description of a utility patent application must include background information, such as a discussion of prior art (older inventions). It is usually not necessary to make a drawing of prior art. If you do, such figures must be numbered in the conventional manner in sequence with the rest of the figures, and the label "Prior Art" must be added next to or below the figure number, such as in Illustration 8.13.

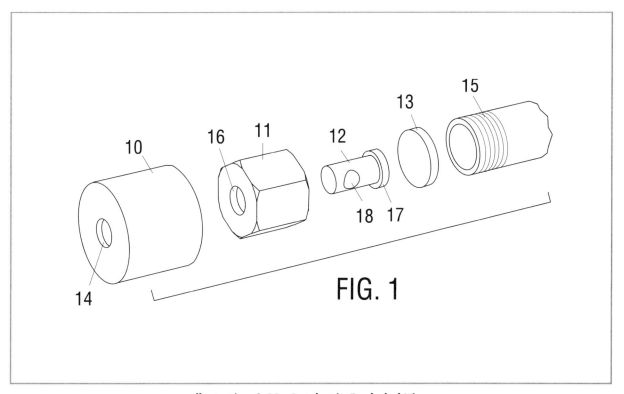

Illustration 8.11—Bracket in Exploded View

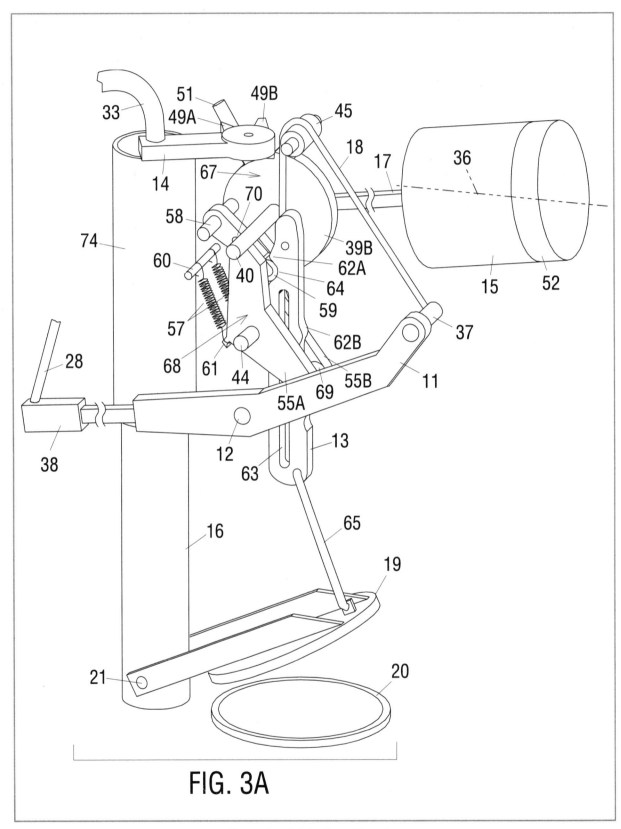

FIG. 3A

Illustration 8.12—Use of Bracket

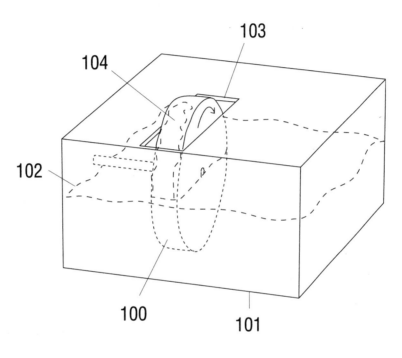

Fig. 1
Prior Art

Illustration 8.13—"Prior Art" Label

D. Reference Numbers

Each part or element of a utility patent drawing mentioned in the written description must be designated by a reference number or letter, although numbers are preferred. They may start with any number, and do not have to be consecutive. Design patent drawings must not include reference numbers or letters.

Reference numbers mentioned in the description of the invention in the specification must appear in the drawings, and reference numbers that appear in the drawings must be mentioned in the description. That is, each number must appear in both the specification and the drawings. Let's take a very simple description as an example: "As shown in Fig. 1, a chair includes a seating surface 10, supporting legs 11, 12, 13, and 14, and a seat back 15." For simplicity, consider it to be the entire description in the application. The same reference numbers, which include 10, 11, 12, 13, 14, and 15, must all be shown in the drawings.

1. Different Parts Must Have Different Numbers

Each designated component of the invention must have a unique reference number—that is, the same number must not be used to designate different parts of the invention. If the same part appears in separate figures, it must be designated with the same number. For simplicity, the same number may be used to designate separate but identical parts, such as bolts. However, if such parts are referred to separately, different numbers should be used to avoid confusion. For example, "A bolt 13 is removable from the top end of a connecting rod 14, while bolts 15 are permanently attached to the lower end of connecting rod 14."

⚠ **Keep your parts and numbers straight.** *Using different numbers for the same part in different figures, or using the same number for different parts, are two of the most common mistakes. Proofread the description and drawings carefully to weed out such errors.*

2. Consecutive Numbers Are Preferable

There is no PTO requirement for reference numbers to appear in any particular order in the written description (specification). Nevertheless, consecutive numbers (for example, 10, 11, 12, 13, etc.), or consecutive odd or even numbers (for example, 11, 13, 15, 17, etc., or 10, 12, 14, 16, etc.), are preferable, because they are easier for a reader to follow than random numbers that do not follow a logical sequence (for example, 24, 12, 43, 11, 33, etc.). Also, the numbers customarily start with at least 10, or a higher number than the highest figure number, to avoid confusion with the figure numbers. Using every other number, as suggested in *Patent It Yourself*, allows intermediate numbers to be added later, and still maintain the sequence. For example, if after writing much of the specification, you want to number a washer mentioned between a bolt 16 and a nut 18, you can give it the number 17. However, if you used consecutive numbers, and need to add a number, simply use the next higher number; don't worry about renumbering everything.

3. Reference Letters

Instead of reference numbers, suitable letters may be—but are not necessarily—used for designating non-tangible elements, such as "A" for air flow, and should be of the English alphabet. Letters may also be used where they are customarily used, such as in electronics. For example, "R1," "R2," etc. may be used for designating resistors; "C1," "C2," etc. may be used for designating capacitors; and so on.

4. Non-English Letters

Non-English letters are allowed where there is customary usage. For example, Greek letters may be used to indicate mathematical and scientific values in formulas and the labels of graphs, and they may also be used as reference characters in drawings.

5. Primed Numbers

Primed numbers or characters—that is, numbers with a mark (') to their upper right—for example, 23', 23", etc.—are not specifically prohibited, but they should be avoided because they tend to be confusing.

6. Letter Suffixes

Reference numbers with letter suffixes may be used if the letter suffix helps explain the part, for example "left wheel 12L and right wheel 12R," or "upper surface 14U and lower surface 14L," etc.

7. Number Size and Style

Reference numbers or letters must be of a simple lettering style or font, and at least 3.2 mm, 1/8" or 14 points in height, as shown in Illustration 8.14. Avoid the very ornate type of lettering used in some old patents. Do not make the reference numbers larger than the required size, otherwise they take up too much room and may make the sheet too crowded for easy reading, or may not fit into small spaces. To further save space, use a thin, horizontally compact style or font, which is typically called "condensed."

An exception to the size requirement is that subscripts, such as the "2" in the chemical formula "CO_2," may be smaller than 3.2 mm.

53

Undesirable: Ornate lettering style or font

53

Better: Simple style or font, but too thick and wide

53

Preferable: Simple lettering style or font

Illustration 8.14—Reference Number Style or Font

When drawing with a computer-aided draft-ing (CAD) program, select the font before applying the reference numbers. This is because if you change the font after the numbers are applied, they will shift position and will no longer line up properly with the lead lines. The amount of shifting depends on the fonts used before and after the change.

A keystroke saver. *Some CAD programs require the font height to be typed in every time the program is used, so frequently typing in "3.2" gets tiresome. It is more convenient to type in just "3"; a 3 mm font is virtually indistinguishable from a 3.2 mm font, so that it is perfectly acceptable.*

8. Size of Descriptive Text

In addition to reference numbers and letters, the size requirement applies to all characters, including lower case letters, used in any descriptive text in the drawings. As stated, all letters must be at least 3.2 mm high. If descriptive text is written in sentence case, such as "Electrical power supply," or title case, such as "Electrical Power Supply," and the shortest lower case letters must be at least 3.2 mm high, then the upper case letters will be even taller and take up too much room. Therefore, it is usually best to use all capital letters, such as "ELECTRICAL POWER SUPPLY", so that no letter is over 3.2 mm tall.

9. Reference Number Positioning

Reference numbers or letters should be positioned a short distance away from the part they are desig-nating, and far enough from other parts and from each other to avoid confusion, as shown in Illus-tration 8.15. They are preferably positioned outside the figure to avoid cluttering it, but they may be positioned inside if necessary to avoid being too

far away. They must never be positioned across any lines of the figure, including hatch lines.

10. Reference Numbers Between Different Embodiments

In an application with multiple embodiments or variations that share common parts, the same refer-ence numbers can be used for the common parts between the embodiments. For example, two embodiments of a hand pump include the same pump cylinder, but different handles. The identical pump cylinders may be designated with the same reference number in all the figures—for example, "10"—but the handles must be designated with different numbers, such as "11" in the figure of the first embodiment, and "12" in the figure of the second embodiment.

E. Lead Lines

A lead line must extend between each reference number and its corresponding part. One end of the lead line should be very close to the reference number without touching it, and the other end must touch the part of the figure being designated, such as in Illustration 8.16. Lead lines must never cross each other.

1. Either Straight or Curved Lines

Lead lines may be straight or curved; a drawing can include both types if you wish. Although straight lines are easier to draw, curved ones can be more easily positioned in tight places, as shown in Illustration 8.16. They should not be too long—that is, the reference numeral should not be positioned unnecessarily far away from the part so as to require a long lead line. They should be positioned at a large angle from adjacent lines of

Correct

Numbers positioned outside figure

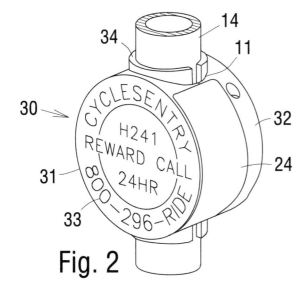

Fig. 2

Correct

Some numbers positioned inside figure
without crossing lines of figure

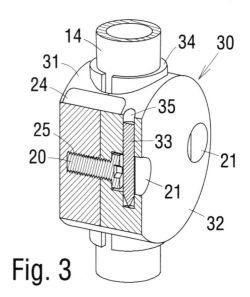

Fig. 3

Incorrect

Numbers positioned over lines of figure

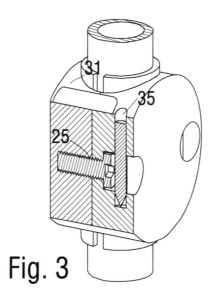

Fig. 3

Illustration 8.15—Reference Number Positioning

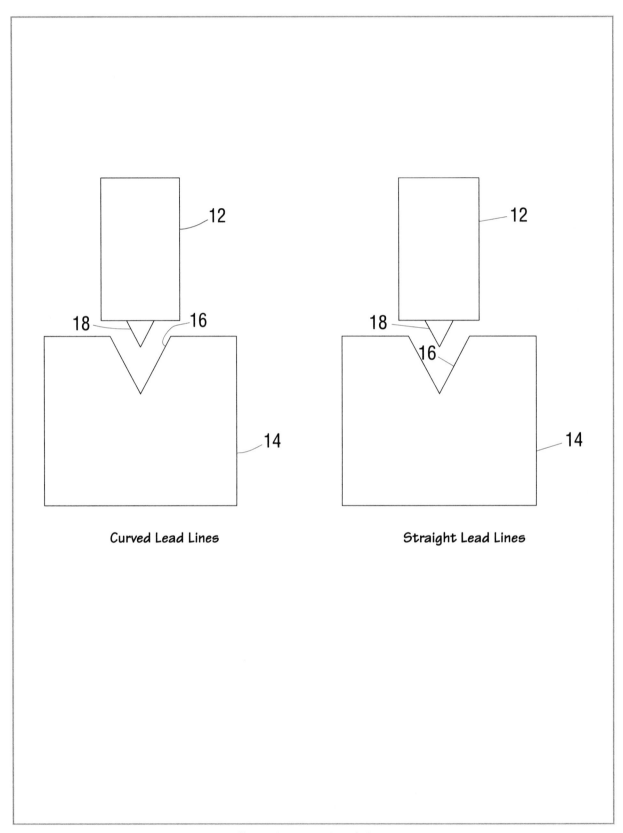

Curved Lead Lines   Straight Lead Lines

Illustration 8.16—Lead Lines

the figure to avoid being confused as lines of the figure. They may cross lines of the figure, but should be positioned so as to cross as few lines as possible to avoid confusion.

💡 *For best appearance, a lead line should be aligned with the imaginary center of the reference number, as shown in Illustration 8.17. The dashed line is solely for showing alignment; it should not appear in the drawing.*

2. Lead Lines Must Not Connect Separate Figures

If an identical part appears in separate figures, it must not be referenced by a single reference numeral with two lead lines. In Illustration 8.18, a badge is shown in different views in Figs. 2 and 3.

The top two illustrations show the incorrect usage of lead lines: the pipes in each figure are connected to the same reference number (14) positioned between the figures. The figures are thus considered to be "connected" by the lead lines, which is improper. The bottom two illustrations show the correct usage of lead lines: the pipes in each figure are connected to their own reference numbers, so that the figures are not connected.

3. Replacing Lead Lines With Underline

If a reference number lies on a surface or cross section it is meant to designate, the lead line may be replaced with an underline for the number, such as parts 11 and 12 in Illustration 8.19. In fact, part 10 may also be underlined instead of being

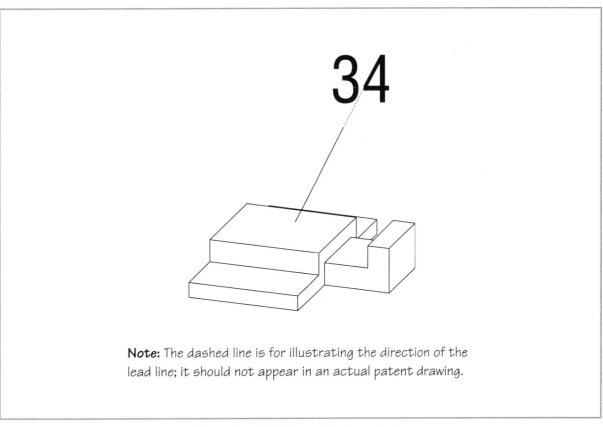

Note: The dashed line is for illustrating the direction of the lead line; it should not appear in an actual patent drawing.

Illustration 8.17—Lead Line Positioning

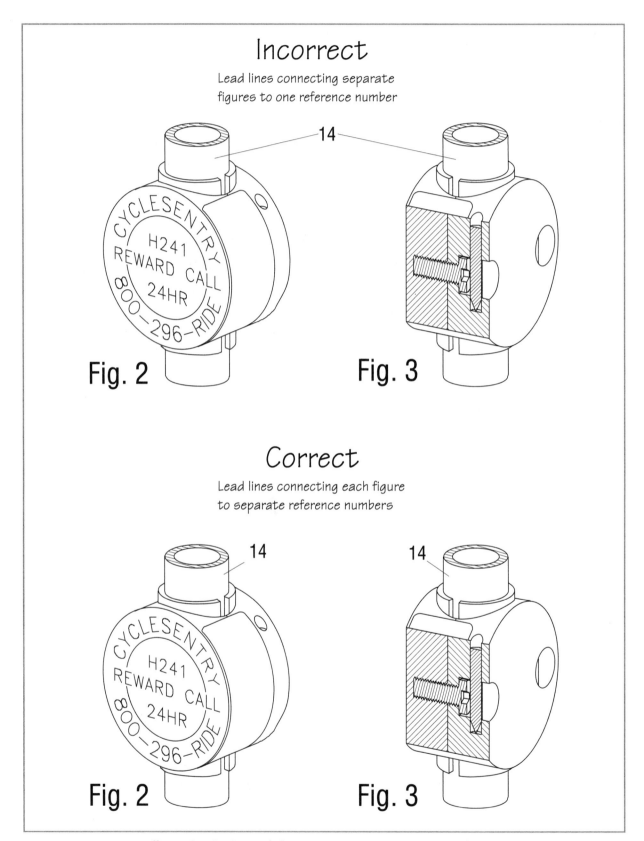

Illustration 8.18—Lead Lines Must Not Connect Separate Figures

connected with a lead line. This method may be advantageous in some situations where space is limited. The hatching in the cross section must be interrupted to allow space for the number. However, this type of numbering may be confusing, so it should be used sparingly.

F. Arrows

Arrows are used in the following situations:

1. To designate a group of parts with a single number, even if each part has its own number, as shown in Illustration 8.20. The arrow should point in the general direction of the group. In the illustration shown, parts 17, 39, 41, 42, and 51 are collectively designated as an assembly 67; and parts 43, 44, 55, and 66 are collectively designated as an assembly 68 (the term "assembly" is chosen arbitrarily). The specification should describe the assembly as including such-and-such constituent parts—for example, "A cam assembly 67 includes arms 17 and 51, a disc 41…"

2. To indicate the plane and direction of a sectional view. (See Chapter 6, Section E, for details.)

3. To show the direction of movement, as shown in Illustration 8.21. The purpose of such an arrow should be mentioned in the specification—for example, "a hinged arm 12 pivoting in the direction indicated by the arrow." If there is more than one arrow in a figure, each arrow should be designated with a different reference number to avoid confusion.

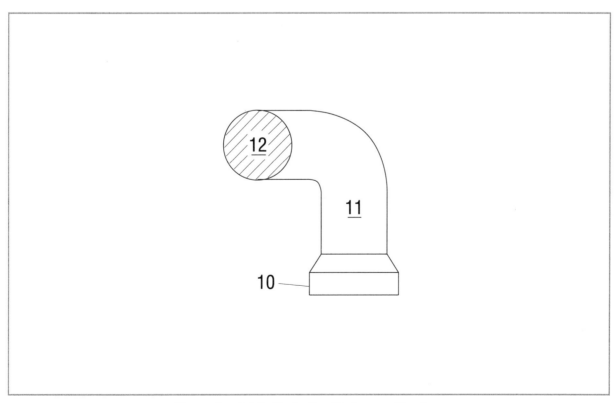

Illustration 8.19—Underlined Reference Numbers

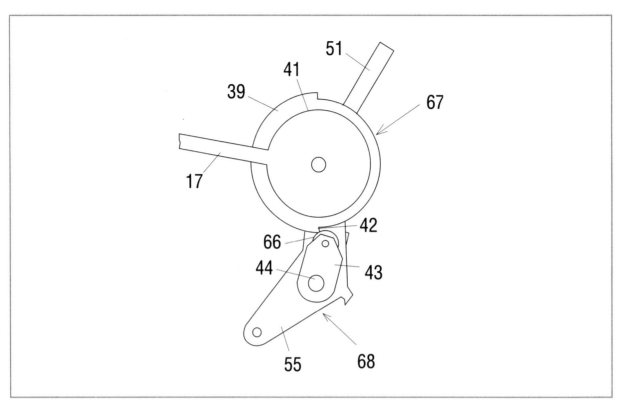

Illustration 8.20—Arrows to Designate Groups of Parts

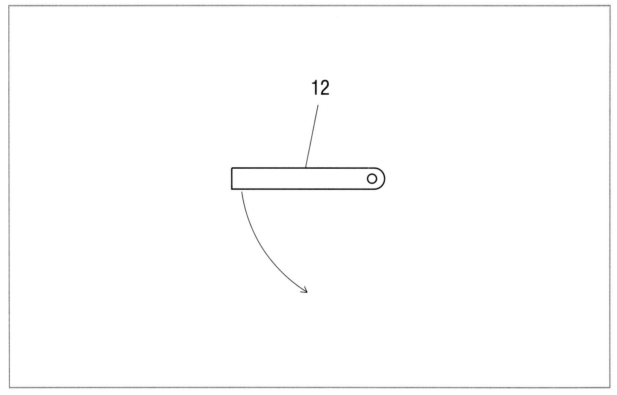

Illustration 8.21—Arrow to Indicate Movement

G. Line Types

Allowable line types are shown in Illustration 8.22.

Solid Line. Used for edge lines and shading lines.

Hidden Line (Dashed). Sometimes it is necessary to show a part hidden behind other parts. Such hidden parts are shown with dashed lines.

Phantom Line (Dash-Dot-Dot-Dash). A component which is not part of the invention may be shown in phantom lines. This is absolutely necessary in design patent drawings. Such a component may be shown in solid lines in utility patent drawings.

Projected Line (Dash-Dot-Dash). Projected lines may be used to show how separated objects fit together, although such lines are not usually necessary. The use of a projected line in an exploded view eliminates the need for a bracket. (See Chapter 6, Section C, for details on exploded views.)

Although technically there are different broken lines (hidden, phantom, and projected) for different applications, the dashed line may be used for any situation that requires a broken line.

H. Character of Lines

Except for color drawings and photographs, all drawings must be done in black lines. The lines must be thick enough to allow photocopying without loss of detail, although the thickness of lines may vary according to their role in a drawing. All lines must be solid black, uniformly thick, and have sharp, smooth edges.

Common Errors: *Lines that are not dense or black enough, not uniformly thick throughout, jagged, or have feathered edges, as shown in Illustration 8.23. These are by far the most common errors found in patent drawings, and are typically made by using the wrong equipment—such as the wrong paper, pen, or computer printer—or by using an improper drafting technique. (See Chapters 2*

and 3 on the proper equipment and techniques for making lines.)

Although all the lines in a drawing may be of the same width, the use of different widths for lines in different roles can greatly improve the legibility and aesthetics of a drawing. (See Chapters 6 and 7 for suggested line widths.)

I. Descriptive Legends

Descriptive text is permitted in a drawing only when it is absolutely necessary. One situation where it is necessary and allowed is within the boxes of flowcharts and block diagrams. (Flowcharts and block diagrams are discussed in Chapter 6, Section I.)

J. Scale of Drawing

The same object should be shown in different figures in the same size if possible. Each figure should be large enough so that all of its essential details are easily comprehensible. Use all of the available space on a sheet to make each figure as large as possible without crowding the figures. Use the whole sheet for a single figure if necessary, as in Illustration 8.24.

Common Error: *Drawing the figures too small, so that the lines are too close together and the features are too small to be easily discerned, as in Illustration 8.25.*

Do not hesitate to enlarge an object in a separate figure for clarity whenever it is necessary. If an object is shown substantially larger in a figure relative to another, the description (specification) should note the different scales to avoid confusion —for example, "mechanism 15 is shown enlarged in Fig. 3 for clarity."

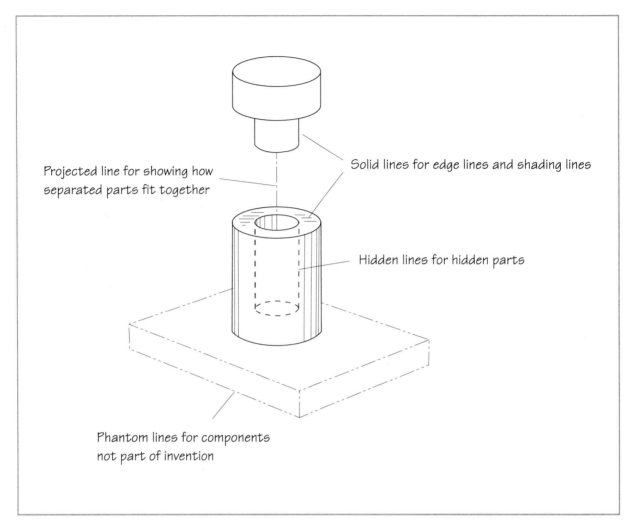

Projected line for showing how separated parts fit together

Solid lines for edge lines and shading lines

Hidden lines for hidden parts

Phantom lines for components not part of invention

Illustration 8.22—Line Types

Not Dense Enough

Not Uniformly Thick

Jagged

Illustration 8.23—Poor Line Quality

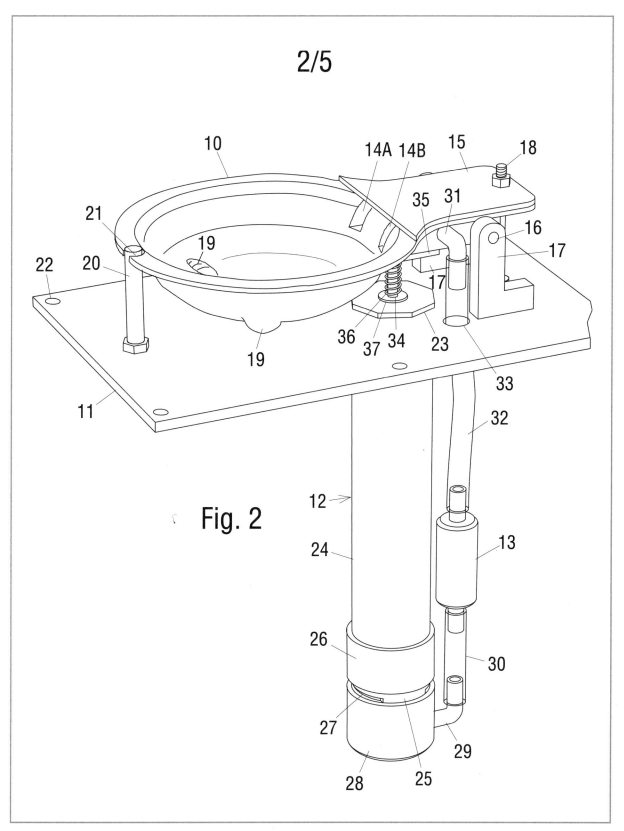

2/5

Fig. 2

Illustration 8.24—Using Full Sheet to Show All Details

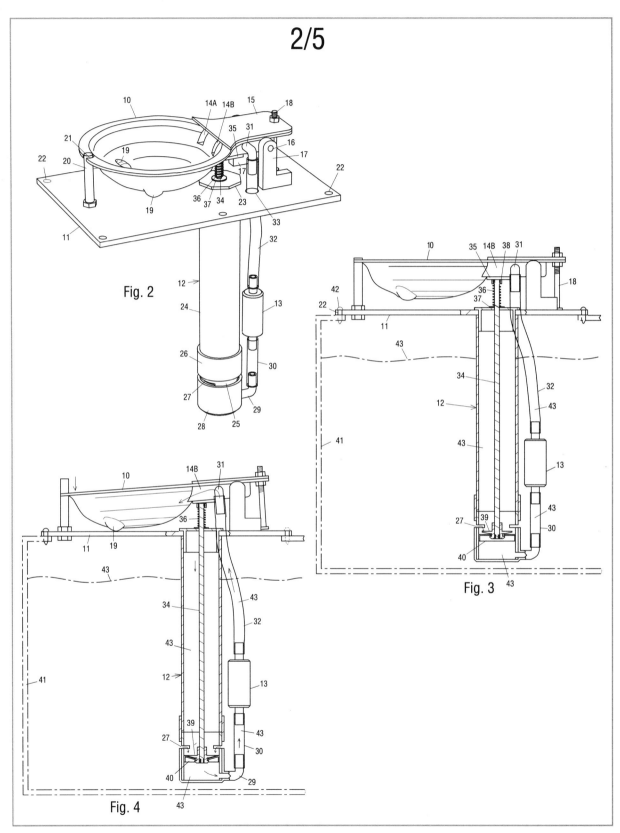

2/5

Fig. 2

Fig. 3

Fig. 4

Illustration 8.25—Figures Too Small and Crowded

If only part of an object is enlarged in a separate figure, the part of interest should be indicated in the original figure by a dashed circle, as shown in Illustration 8.26, so as to indicate its location within the whole object. However, the PTO rarely objects to the lack of such a dashed circle. Different figure numbers must be used to designate the separate figures.

K. Copyright or Mask Work Notice

You may be able to obtain a copyright as well as a utility patent for some inventions, notably objects with aesthetic features and functional parts. Refer to *Patent It Yourself*, Chapter 1, for more information on copyright.

If you are applying for a utility patent for an invention that is also covered by copyright, you may include a copyright notice in the patent drawing to publicize it. Illustration 8.27 shows a wind vane that uses an electronic position encoder 12, and an ornamental, cat-shaped vane 13. The whole invention (a position encoder attached to a wind vane) may be covered by a utility patent, while the cat-shaped vane alone may be covered by a copyright. The proper format of the copyright notice is "© year of copyright your name." The following paragraph must be included at the beginning—preferably as the first paragraph—of the specification:

"A portion of the disclosure of this patent document contains material which is subject to copyright protection. The copyright owner has no objection to the facsimile reproduction by any one of the patent disclosure, as it appears in the Patent and Trademark Office patent files or records, but otherwise reserves all copyright rights whatsoever."

The copyright notice must be placed below the copyrighted design within the margins—that is, within the sight—of the sheet. The font or lettering must be between 3.2 mm (1/8" or 14 points) and 6.3 mm (1/4" or 28 points) high.

A special category of copyright is known as a "mask work," which is a mask used in the making of integrated circuits. If your invention is such a mask, substitute "mask work" for "copyright" in the above paragraph. The proper format for a mask work notice is "*M* your name." For example "*M* John Smith."

If the shape of your invention is considered a trademark (brand name, brand symbol), such as the PhotoMat huts in shopping center parking lots, it would be wise to indicate this in the specification (description)—for example, "Shape of building in Fig. 1 is applicant's trademark for a photo finishing service."

L. Security Markings

All applications filed in the PTO are screened for subject matter which might impact our national security. If you have invented, for example, a new doomsday weapon, and the PTO and other federal agencies determine that publicizing your invention by granting a patent will be detrimental to our national security, a secrecy order will be placed on your patent application. The PTO will notify you, and the grant of your patent will be withheld for as long as national security requires.

If your invention is subject to a secrecy order, or any other security classification (for example, you received a secrecy order from the PTO after its security review, or you are working under a government contract, which involves confidential, secret, top secret, or "Q" material), you may place authorized security markings—such as NATO (North Atlantic Treaty Organization), TS (top secret), S (secret), or C (confidential)—in the middle of the upper margin of a sheet, as shown in Illustration 8.28.

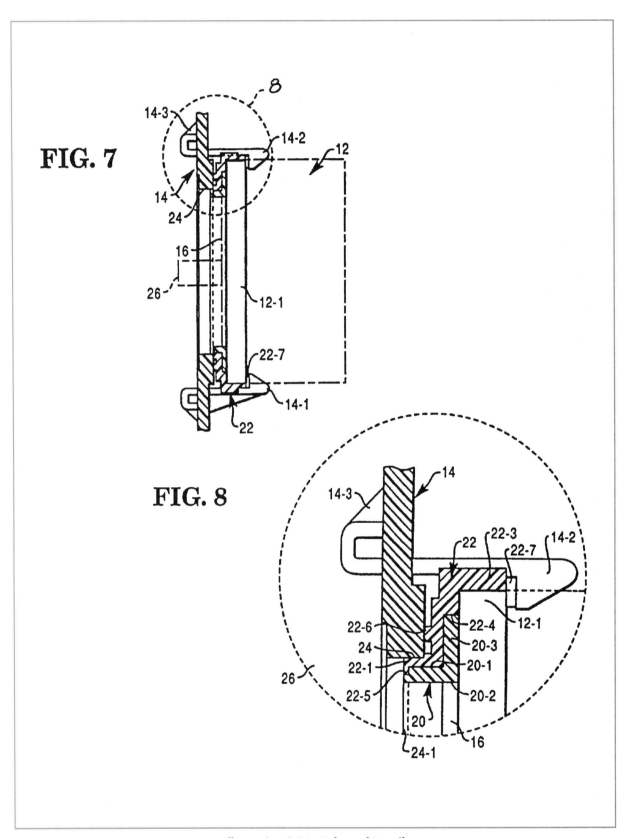

FIG. 7

FIG. 8

Illustration 8.26—Enlarged Detail

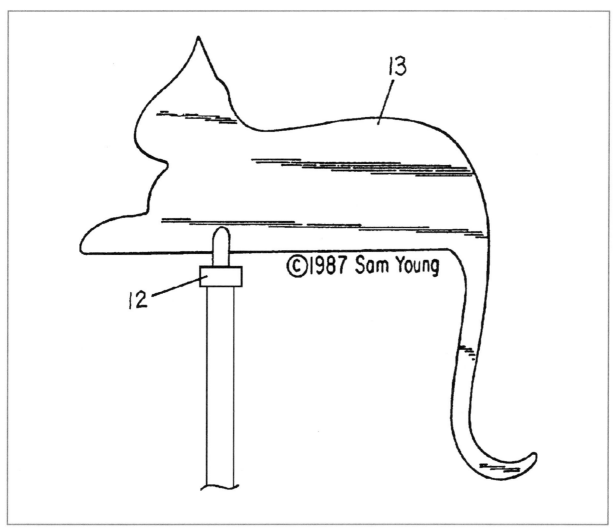

Illustration 8.27—Copyright Notice

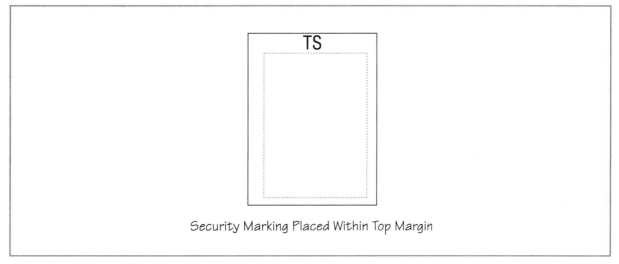

Illustration 8.28—Security Marking

M. Corrections

Ink drawings may be corrected by erasure or white masking fluid, such as White Out®, as long as erased lines are invisible and the masking fluid is durable, so that it will not crack or flake off. If you use an eraser, be careful when drawing new lines over erased areas, which may become roughened, causing the ink to bleed or feather. If you prefer a masking fluid, select one with a pen-type applicator and avoid those with a brush applicator, which tend to dry out quickly and make application difficult. Note that white masking fluid dissolves some inks, which may darken the fluid. Such corrections are not acceptable.

Corrections to a drawing sheet can be avoided if the drawing is made with CAD: the drawing can be corrected on the computer, and a new sheet printed.

N. Prohibited Elements

The following elements cannot appear on patent drawings:

1. Indications regarding the scale of the figures, such as "Actual Size," "Scale 1:2," etc. This is because the PTO often reduces drawings in size when printing them as part of a patent, so scale indications became inaccurate. Such indications should not appear in the description (specification) either.

2. Expressions or drawings contrary to morality or public order, such as explicit sexual and violent images (unless necessary to show the invention), profanity, etc.

3. Trademarks or service marks, such as Coca Cola®, AT&T®, etc., unless you prove that you have a proprietary interest in the mark.

4. Any statement or other matter obviously irrelevant or unnecessary under the circumstances.

5. Descriptive legends (text)—such as "Electrical diagram of widget", "on/off switch", "serrated surface", etc.—except one or a few words when such words are indispensable—for example, within the individual boxes of flow charts and schematics.

6. Center lines depicting the center or axis of circular parts. However, this rule is rarely enforced.

7. Brackets or circles surrounding reference numerals. Although not specifically prohibited, brackets should not surround reference numerals in the description (specification) either.

8. Any lines connecting separate figures (figures with different figure numbers), except for electrical waveforms.

9. Solid black shading areas, except when used to represent bar graphs, or black color in drawings meant to depict color as a distinguishing feature of an invention.

O. Identification Information

Optional identification information may be placed on the back of each sheet. Although it is optional, such information may become useful if your drawings are separated from your file at the PTO. Such information should include the following:

1. Title of Invention.
2. Name of Inventor.
3. Application serial number.
4. Group art unit.

The lettering must not show through on the front of the drawing, so write them lightly with a pencil near the edge of each sheet, or print them in light gray with a laser printer. Items 3 and 4 are known only after the application has been filed and you have received a filing receipt, so they only have to be used when filing corrected drawings, such as in response to Office Actions. (See Chapter 9, Responding to Office Actions, for details on filing corrected drawings.) ■

Responding to Office Actions

fter a regular patent application is filed, the first communication from the PTO you will receive is a blue Filing Receipt. The waiting time for the next communication varies between about two months to over a year. If you have done a superb job in preparing the application, and if the examiner does not find close prior art (older inventions, typically found in existing patents), you may receive a Notice of Allowance and Issue Fee Due, which indicates that the application has been approved. However, in most cases you will receive an Office Action, which is a multi-page letter from an examiner who examined your application. The Office Action indicates that the application is not approved for one or more reasons. This chapter details how to deal with Office Actions that raise drawing issues.

A. Objections and Rejections

The drawings in each regular patent application are examined by two branches of the PTO: the Examining Branch and the Drawing Review Branch. The results of the examination are included in the Office Action, which details one or more objections to and/or rejections of the application under one or more statutes (laws) and/or rules. The drawings in a provisional patent application are not examined; see *Patent If Yourself* for details on provisional applications.

A rejection is applied to the claims (the part of an application that defines an invention in legal terms); whereas an objection is typically applied to other aspects of an application, such as incorrect paper size, incorrect line spacing in the specification, misspellings, mismatched reference numbers between the description and drawings, unclear description, inadequate detail in the drawings, etc. All objections and rejections must be overcome by fixing the cited defects, or successfully arguing against them. Otherwise, a patent will not be granted.

Each branch may independently object to the drawings for non-compliance with certain statutes and rules. For example, the Drawing Review Branch may object to certain informal matters—such as rough lines or improper margins—and the Examining Branch may object to more substantive matters—such as a wrong reference number or, worse, failure to illustrate a part specified in the claims. New, corrected drawings must be filed to overcome all objections. Refer to *Patent It Yourself*, Chapter 13, for how to handle objections and rejections relating to the specification and claims, and general information on how to prepare a response or amendment to the Office Action.

There are many possible drawing-related objections. This chapter details how to handle the most common ones. An Office Action always gives reasons for any objection, so you will understand it even if you get one that is not covered here. If you do not understand an objection, you may call your examiner or drafting reviewer to ask for clarification; their helpfulness varies from person to person. The phone number of the examiner is provided on the last page of the Office Action, and the phone number of the Drafting Branch is provided on the Notice of Draftsperson's Patent Drawing Review (discussed in Section H, below), which is included with the Office Action. You should feel free to call these numbers if you have questions, but obviously you should try not to antagonize the person on the other end of the line, as that will hurt your chances.

B. Reading the Statute and Rule Numbers

An Office Action always cites specific laws and rules for the objections or rejections, such as 35 U.S.C. § 102; 37 C.F.R. § 1.83, etc. "35 U.S.C." means Title 35 of the United States Code, which is the patent statutes. "37 C.F.R." means Title 37 of the Code of Federal Regulations, which is a compilation of the patent rules. "§" is the section symbol, and the number following it is the specific section of the statute or rule that is being applied to the application. Statues are laws, which are written in relatively broad terms, whereas rules are written to expound on the laws and are more specific.

C. Two Sets of Requirements

Patent drawings, in conjunction with the written description, are examined by the Examining Branch—specifically, by an examiner—to see whether they are detailed enough to enable someone skilled in the art (pertinent field) to make and use the invention. This is known as the "enablement requirement."

The drawings are also examined by the Drawing Review Branch—specifically, by a draftsperson—for compliance with the general drawing standards, which are discussed in Chapter 8, General Standards. The drawing standards are essentially concerned with the aesthetic appearance of the drawings, and are separate from the enablement requirement. A Notice of Draftsperson's Patent Drawing Review is usually included in the first Office Action for applications that include drawings. It indicates whether the drawings are acceptable and, if they are not, details why they are not acceptable. Drawings that do not comply with all the general standards are known as having "informalities."

The two branches review the drawings independently, so that they may be acceptable to one branch but not the other.

D. Objection or Rejection Under 35 U.S.C. § 112

As already stated, an Office Action details objections or rejections to an application under various statutes and rules. An objection or rejection may be made under 35 U.S.C. § 112, and reads as follows:

"The specification is objected to under 35 U.S.C. § 112, first paragraph, as failing to provide an adequate written description of the invention, and to provide an enabling disclosure." ("Enabling disclosure" means information, including written description and drawings, that are detailed and clear enough to enable someone skilled in the pertinent field to make and use the invention.)

or

"The claim is rejected under 35 U.S.C. § 112, first paragraph, as the claimed invention is not described in such full, clear, concise and exact terms as to enable any person skilled in the art to make and use the same." ("Art," in this context, means the field of the invention.)

Such an objection or rejection may be applied to utility or design patent applications.

1. When Applied to Utility Patent Application

When an objection or rejection under 35 U.S.C. § 112 is made in a utility patent application, the drawings may not be mentioned. However, they are usually also affected, because the drawings and description cooperate to provide the enabling disclosure.

Such an objection or rejection may be overcome if you can argue that the disclosure is in fact enabling—that is, it describes and/or shows the invention clearly enough to enable one skilled in the field to make and use the invention. (Refer to *Patent It Yourself,* Chapter 13, for how to make such arguments.) Otherwise, additional description and/or drawings must be added by filing a continuation-in-part (CIP) application. See Section I2, below, and *Patent It Yourself,* Chapter 14, for details on filing a CIP application.

EXAMPLE 1: An application for a stair-climbing, motorized wheelchair describes and shows in detail the mechanism for climbing stairs. The application includes a circuit diagram which does not show a source of electrical power. An examiner objected to the specification and drawings, and rejected the claims for failure to provide an enabling disclosure, because a source of electrical power is neither described nor shown. The objection and rejection can be overcome by reasonably arguing that anyone skilled in the field, such as an engineer, knows to provide a suitable source of power—such as a battery—and that the power supply is customarily omitted in circuit diagrams.

EXAMPLE 2: An application describes a self-healing tire that automatically seals punctures, but does not describe how it is accomplished. The drawings only show the exterior of the tire. An examiner objects to the specification and drawings, and rejects the claims for failure to provide an enabling disclosure, because the application neither describes nor shows how to make the invention. In this case, the objection and rejection can only be overcome by filing a CIP application to provide details on how the self-healing works.

2. When Applied to Design Patent Application

An objection or rejection under 35 U.S.C. § 112 may also be applied to a design patent application, because the precise or complete appearance of the invention is unclear. The Office Action will specify why the drawings are inadequate—for example, they lack shading, are roughly drawn, do not show all sides of the invention, etc. It is usually difficult to successfully argue that the drawings are adequate in this situation, so the only response is to fix the drawings to provide the additional information. However, you may add features to a drawing *only* if the information being provided is already shown in some of the other drawings as filed. If the information being added is not shown in any way in the original drawings, you must file a CIP application. See Section I2, below, and *Patent It Yourself*, Chapter 14, for details on CIP applications.

EXAMPLE: The original drawings of an application include a front perspective view of a chair. The figure lacks shading, so that the surface contours of the seat and back rest are not clear. An examiner objects to the figure for lacking shading. Fortunately, the drawings also include a side sectional view that clearly shows the contours of those parts. Surface shading may thus be added to the perspective view to overcome the objection without adding new matter, because the contours are already shown in the side sectional view. Additional drawings may also be added to show the chair from different angles if they do not reveal any features that are not in the original drawings. However, if the original drawings do not include a side sectional view, so that the contours of the seat and back rest are not clearly shown in any figure, then the only way to overcome the objection is to file a CIP application to provide the sectional view or shading.

E. Objection Under 37 C.F.R. § 1.83(a) for Failure to Show Claimed Feature

In utility or design patent applications, an objection may be made under 37 C.F.R. § 1.83(a). Such an objection reads as follows:

"The drawings are objected to under 37 C.F.R. § 1.83(a). The drawings must show every feature of the invention specified in the claims. Therefore, the [element] in claim [x] must be shown or the feature deleted from the claim. No new matter should be entered."

Such an objection is applied when a claim recites (mentions) a certain element (feature), but such element is not shown in the drawings. The element may be added to the drawings without introducing new matter, provided that what is added to the drawings is no more detailed than what is recited in the specification or claim. That is, as discussed above, unless the element is very simple or readily understood by one skilled in the art—such as a hinge, a switch, a power supply, etc.—the description or claims must be detailed enough to support the changes to the drawings. Otherwise, you must delete the element from the claim, or file a CIP application to add the element to the drawings. (See Section I2, below, and *Patent It Yourself*, Chapter 14, for details on CIP applications.) Even if the feature may be properly added to the drawings, you may choose to delete it from the claim if it is not important, so as to avoid having to make a new drawing.

F. Objection Under 37 C.F.R. § 1.84(p)(4) for Improper Reference Numbers

In utility patent applications, an objection may be made under 37 C.F.R. § 1.84(p)(4). Such an objection reads as follows:

"The drawings are objected to under 37 C.F.R. § 1.84(p)(4). The same part of an invention appearing in more than one view of the drawing must always be designated by the same reference character, and the same reference character must never be used to designate different parts."

Such an objection is applied when there is a mismatch in the reference numbers between the different figures. It may be overcome by changing the numbers in the drawings to ensure that the same parts in different figures have the same number, and that different parts never share the same number. Having a list of reference numbers in the specification may help you keep the numbers straight and avoid this objection.

EXAMPLE 1: The objection is made because the number 12 is used to designate a bracket in Fig. 1, but the number 18 is used to desig-

nate the same bracket in Fig. 2—that is, different numbers are used to designate the same element. The objection can be overcome by changing 12 to 18 in Fig. 1, or changing 18 to 12 in Fig. 2, so that the bracket is designated with the same number in both figures. The bracket must be also properly numbered in the written description to correspond with the drawings.

EXAMPLE 2: The objection is made because a lever and a button are both designated with the number 24 in the drawings—that is, the same number is used to designate different elements. The objection can be overcome by changing the drawings so that the lever and the button are designated with different numbers. Again, the elements must also be numbered in the written description to correspond with the drawings.

G. Objection Under 37 C.F.R. § 1.84(p)(5) for Missing Reference Numbers

In utility patent applications, an objection may be made under 37 C.F.R. § 1.84(p)(5). Such an objection reads as follows:

"The drawings are objected to under 37 C.F.R. § 1.84(p)(5). Reference characters not mentioned in the description shall not appear in the drawings. Reference characters mentioned in the description must appear in the drawings."

Such an objection is applied when there is a mismatch between the reference numbers in the drawings and those in the written description, so that a reference number is "missing." This objection may be overcome by making changes to either the drawings or description to ensure that each reference number appears in both the description and drawings, and never just one or the other. Again, including a list of reference numbers in the specification may help you keep the numbers straight and avoid this objection.

EXAMPLE 1: The objection is made because the number 32 is used to designate a cylinder in the drawings, but the number 27 is used to designate the cylinder in the written description—that is, different numbers are used in the drawings and the written description to designate the same element. The objection can be overcome by changing 32 to 27 in the drawings, or changing 27 to 32 in the written description, so that the cylinder is designated with the same number in both the drawings and written description.

EXAMPLE 2: The objection is made because a power cord is designated with the number 19 in the drawings, but the same number does not appear in the written description. The objection can be overcome by either deleting the number 19 from the drawings, or adding the number 19 to the written description to designate the power cord, so that the number 19 is used to designate the same element in both the drawings and the written description.

EXAMPLE 3: The objection is made because a socket is designated with the number 36 in the written description, but the same number does not appear in the drawings. The objection can be overcome by either deleting the number 36 from the written description, or adding the number 36 to the drawings to designate the socket, so that the number 36 is used to designate the socket in both the drawings and the written description.

H. Notice of Draftsperson's Patent Drawing Review

A Notice of Draftsperson's Patent Drawing Review, as shown in Illustration 9.1, is usually included in the first Office Action for all applications with drawings. It is issued by the Drawing Review Branch. You will get a well-deserved sense of satisfaction if part A on the top left of the Notice is checked to indicate that the drawings are acceptable.

If part B is checked, the drawings have been determined to be out of compliance with one or more rules. As shown in Illustration 9.1, the specific objections are detailed in categories 1 to 17, each of which has several related items. The probable cause and remedy for each objection is discussed below.

1. Drawings. 37 CFR 1.84(a): Acceptable categories of drawings: Black ink. Color.

__ **Not black solid lines. Fig(s)_____**

Probable cause of objection: The drawings were done with a pencil or other non-black writing instrument.

Remedy: Redo the drawings in solid black lines with the proper pen or printer. (See Chapter 8, Section H, and Chapters 2 and 3.)

__ **Color drawings are not acceptable until petition is granted.**

Probable clause of objection: Color drawings or color photographs have been submitted without filing the required petition.

Remedy: File a petition, or replace them with black line drawings. If a petition has already been filed, the drawings will be accepted when the

Form PTO 948 (Rev. 10-93) U.S. DEPARTMENT OF COMMERCE - Patent and Trademark Office Application No. _____

NOTICE OF DRAFTSPERSON'S PATENT DRAWING REVIEW

PTO Draftpersons review all originally filed drawings regardless of whether they are designated as formal or informal. Additionally, patent Examiners will review the drawings for compliance with the regulations. Direct telephone inquiries concerning this review to the Drawing Review Branch, 703-305-8404.

The drawings filed (insert date)_____, are
A.____ not objected to by the Draftsperson under 37 CFR 1.84 or 1.152.
B.____ objected to by the Draftsperson under 37 CFR 1.84 or 1.152 as indicated below. The Examiner will require submission of new, corrected drawings when necessary. Corrected drawings must be submitted according to the instructions on the back of this Notice.

1. DRAWINGS. 37 CFR 1.84(a): Acceptable categories of drawings: Black ink. Color.
____ Not black solid lines. Fig(s)_____
____ Color drawings are not acceptable until petition is granted.

2. PHOTOGRAPHS. 37 CFR 1.84(b)
____ Photographs are not acceptable until petition is granted.

3. GRAPHIC FORMS. 37 CFR 1.84 (d)
____ Chemical or mathematical formula not labeled as separate figure. Fig(s)_____
____ Group of waveforms not presented as a single figure, using common vertical axis with time extending along horizontal axis. Fig(s)_____
____ Individuals waveform not identified with a separate letter designation adjacent to the vertical axis. Fig(s)_____

4. TYPE OF PAPER. 37 CFR 1.84(e)
____ Paper not flexible, strong, white, smooth, nonshiny, and durable. Sheet(s)_____
____ Erasures, alterations, overwritings, interlineations, cracks, creases, and folds not allowed. Sheet(s)_____

5. SIZE OF PAPER. 37 CFR 1.84(f): Acceptable paper sizes:
21.6 cm. by 35.6 cm. (8 1/2 by 14 inches)
21.6 cm. by 33.1 cm. (8 1/2 by 13 inches)
21.6 cm. by 27.9 cm. (8 1/2 by 11 inches)
21.0 cm. by 29.7 cm. (DIN size A4)
____ All drawing sheets not the same size. Sheet(s)_____
____ Drawing sheet not an acceptable size. Sheet(s)_____

6. MARGINS. 37 CFR 1.84(g): Acceptable margins:

Paper size			
21.6 cm. X 35.6 cm. (8 1/2 X 14 inches)	21.6 cm X 33.1 cm. (8 1/2 X 13 inches)	21.6 cm. X 27.9 cm. (8 1/2 X 11 inches)	21 cm. X 29.7 cm. (DIN Size A4)
T 5.1 cm. (2")	2.5 cm. (1")	2.5 cm. (1")	2.5cm.
L .64 cm. (1/4")	.64 cm. (1/4")	.64 cm. (1/4")	2.5 cm.
R .64 cm. (1/4")	.64 cm. (1/4")	.64 cm. (1/4")	1.5 cm.
B .64 cm. (1/4")	.64 cm. (1/4")	.64 cm. (1/4")	1.0 cm.

Margins do not conform to chart above. Sheet(s)_____
____Top (T) ____ Left (L) ____Right (R) ____Bottom (B)

7. VIEWS. 37 CFR 1.84(h)
REMINDER: Specification may require revision to correspond to drawing changes.
____ All views not grouped together. Fig(s)_____
____ Views connected by projection lines. Fig(s)_____
____ Views contain center lines. Fig(s)_____
Partial views. 37 CFR 1.84(h)(2)
____ Separate sheets not linked edge to edge. Fig(s)_____
____ View and enlarged view not labeled separately. Fig(s)_____
____ Long view relationship between different parts not clear and unambiguous. 37 CFR 1.84(h)(2)(ii) Fig(s)_____
Sectional views. 37 CFR 1.84(h)(3)
____ Hatching not indicated for sectional portions of an object. Fig(s)_____
____ Hatching of regularly spaced oblique parallel lines not spaced sufficiently. Fig(s)_____
____ Hatching not at substantial angle to surrounding axes or principal lines. Fig(s)_____
____ Cross section not drawn same as view with parts in cross section with regularly spaced parallel oblique strokes. Fig(s)_____
____ Hatching of juxtaposed different elements not angled in a different way. Fig(s)_____
Alternate position. 37 CFR 1.84(h)(4)
____ A separate view required for a moved position. Fig(s)_____

Modified forms. 37 CFR 1.84(h)(5)
____ Modified forms of construction must be shown in separate views. Fig(s)_____

8. ARRANGEMENT OF VIEWS. 37 CFR 1.84(i)
____ View placed upon another view or within outline of another. Fig(s)_____
____ Words do not appear in a horizontal, left-to-right fashion when page is either upright or turned so that the top becomes the right side, except for graphs. Fig(s)_____

9. SCALE. 37 CFR 1.84(k)
____ Scale not large enough to show mechanism without crowding when drawing is reduced in size to two-thirds in reproduction. Fig(s)_____
____ Indication such as "actual size" or "scale 1/2" not permitted. Fig(s)_____
____ Elements of same view not in proportion to each other. Fig(s)_____

10. CHARACTER OF LINES, NUMBERS, & LETTERS. 37 CFR 1.84(l)
____ Lines, numbers & letters not uniformly thick and well defined, clean, durable, and black (except for color drawings). Fig(s)_____

11. SHADING. 37 CFR 1.84(m)
____ Shading used for other than shape of spherical, cylindrical, and conical elements of an object, or for flat parts. Fig(s)_____
____ Solid black shading areas not permitted. Fig(s)_____

12. NUMBERS, LETTERS, & REFERENCE CHARACTERS. 37 CFR 1.84(p)
____ Numbers and reference characters not plain and legible. 37 CFR 1.84(p)(l) Fig(s)_____
____ Numbers and reference characters used in conjuction with brackets, inverted commas, or enclosed within outlines. 37 CFR 1.84(p)(l) Fig(s)_____
____ Numbers and reference characters not oriented in same direction as the view. 37 CFR 1.84(p)(l) Fig(s)_____
____ English alphabet not used. 37 CFR 1.84(p)(2) Fig(s)_____
____ Numbers, letters, and reference characters do not measure at least .32 cm. (1/8 inch) in height. 37 CFR(p)(3) Fig(s)_____

13. LEAD LINES. 37 CFR 1.84(q)
____ Lead lines cross each other. Fig(s)_____
____ Lead lines missing. Fig(s)_____
____ Lead lines not as short as possible. Fig(s)_____

14. NUMBERING OF SHEETS OF DRAWINGS. 37 CFR 1.84(t)
____ Number appears in top margin. Fig(s)_____
____ Number not larger than reference characters. Fig(s)_____
____ Sheets not numbered consecutively, and in Arabic numerals, beginning with number 1. Sheet(s)_____

15. NUMBER OF VIEWS. 37 CFR 1.84(u)
____ Views not numbered consecutively, and in Arabic numerals, beginning with number 1. Fig(s)_____
____ View numbers not preceded by the abbreviation Fig. Fig(s)_____
____ Single view contains a view number and the abbreviation Fig.
____ Numbers not larger than reference characters. Fig(s)_____

16. CORRECTIONS. 37 CFR 1.84(w)
____ Corrections not durable and permanent. Fig(s)_____

17. DESIGN DRAWING. 37 CFR 1.152
____ Surface shading shown not appropriate. Fig(s)_____
____ Solid black shading not used for color contrast. Fig(s)_____

ATTACHMENT TO PAPER NO._____   REVIEWER_____   DATE_____

A-6-3

Illustration 9.1—Notice of Draftsperson's Patent Drawing Review

Although the PTO form still refers to four sizes of paper, new rules only allow two sizes—8 1/2" X 11" and A4.

petition is granted, provided that there are no other objections. If the petition is denied, black line drawings must be substituted for them. (See Chapter 8, Section B, for details on petitions.)

2. Photographs. 37 CFR 1.84(b)

__ **Photographs are not acceptable until petition is granted.**

Probable cause of objection: Photographs have been submitted without filing the required petition.

Remedy: File a petition, or replace them with black line drawings. If a petition has been already filed, the drawings will be accepted when the petition is granted, provided that there are no other objections. If the petition is denied, black line drawings must be substituted for them. (See Chapter 8, Section B.)

3. Graphic Forms. 37 CFR 1.84(d)

__ **Chemical or mathematical formula not labeled as separate figure. Fig(s) _____**

Probable cause of objection: Several formulas have been labeled with a single figure number.

Remedy: Label each formula with a different figure number.

__ **Group of waveforms not presented as a single figure, using common vertical axis with time extending along horizontal axis. Fig(s) _____**

Probable cause of objection: Several electrical waveforms have been presented as separate, unconnected figures.

Remedy: Group the waveforms together, so that they share a common vertical axis on the left

and extend to the right as they proceed. (See Chapter 6, Section F5.)

__ **Individual waveform not identified with a separate letter designation adjacent to the vertical axis. Fig(s) _____**

Probable cause of objection: Several waveforms share a common vertical axis, but they are labeled with a single figure number.

Remedy: Label each waveform with a different figure number next to the vertical axis. (See Chapter 6, Section F5.)

4. Type of Paper. 37 CFR 1.84(e)

__ **Paper not flexible, strong, white, smooth, nonshiny, and durable. Sheet(s) _____**

Probable cause of objection: The drawings have been done on poor paper—perhaps yellow notepad paper, grid paper, napkin, wrinkled paper, easily erasable paper, translucent paper, etc.

Remedy: Redo, photocopy, or reprint the drawings on all white, non-shiny, and smooth paper, such as copier or laser printer paper. (See Chapter 8, Section A.)

__ **Erasures, alterations, overwritings, interlineations, cracks, creases, and folds not allowed. Sheet(s) _____**

Probable cause of objection: The drawings include incomplete erasures, graphic elements, or text that has been overwritten with heavy lines, or the paper is cracked, creased, or folded.

Remedy: Redo, photocopy, or reprint the drawings to ensure that there are no incomplete erasures and overwriting, and that the paper is not cracked, creased, or folded. (See Chapter 8, Sections A and M.)

5. Size of Paper. 37 CFR 1.84(f): Acceptable paper sizes:

__ **All drawing sheets not the same size. Sheet(s) _____**

Probable cause of objection: The specific drawing sheets indicated are of a different size from the rest.

Remedy: Resubmit the ones indicated to match the size of the remaining sheets. (See Chapter 8, Section A4.)

__ **Drawing sheet not an acceptable size. Sheet(s) _____**

Probable cause of objection: The drawing sheets are not one of the acceptable sizes.

Remedy: Redo, reprint, or photocopy the drawings onto paper of an acceptable size. (See Chapter 8, Section A4.)

6. Margins. 37 CFR 1.84(g): Acceptable margins:

Margins do not conform to chart above. Sheet(s) _____
__ **Top (T) __ Left (L) __ Right (R) __ Bottom (B)**

Probable cause of objection: Drawing figures have intruded into the specified margin.

Remedy: Move the figures out of the specified margin. (See Chapter 8, Section A4.) We have received this objection even when the figures are clear of the margins by several millimeters. The PTO drawing reviewers seem to use different rulers than we do, or they may make sloppy measurements. If this happens to you, just move the figures even farther away from the edge. (See Section J, below, for techniques on making changes.) Alternatively, if you don't mind dealing with a large bureaucracy, you may call the Drafting Review Branch and ask to have your drawing margins measured again.

7. Views. 37 CFR 1.84(h)

__ **All views not grouped together. Fig(s)_____**

Probable cause of objection: Consecutive figures are distributed randomly across separate sheets—for example, sheet 1 includes Figs. 3, 4, and 6; sheet 2 includes Figs. 5, 2, and 1, etc.

Remedy: Regroup the figures so that each sheet includes consecutive figures. For example, put Figs. 1, 2, and 3 on sheet 1; Figs. 4, 5, and 6 on sheet 2; etc., using the techniques discussed in Section I, below. (See also Chapter 8, Section C2.)

__ **Views connected by projection lines. Fig(s)_____**

Probable cause of objection: Dashed lines extending between figures with different figure numbers.

Remedy: Delete the dashed lines. (See Chapter 8, Section C1.)

__ **Views contain center lines. Fig(s)_____**

Probable cause of objection: Center lines (crosses indicating the center of round objects) are present in the specified figures.

Remedy: Delete the center lines. (See also Chapter 8, Section N.)

Partial views. 37 CFR 1.84(h)(2)

__ **Separate sheets not linked edge to edge. Fig(s)_____**

Probable cause of objection: The portions of a large drawing spread across several sheets cannot be tiled (positioned) together to display the whole drawing in a clear manner.

Remedy: Rearrange the portions so that they may be tiled to form the whole drawing neatly, and without blocking any part of it. (See Chapter 6, Section C3.)

__ View and enlarged view not labeled separately. Fig(s)_____

Probable cause of objection: Two figures, one showing an enlarged portion of another, share the same figure number.

Remedy: Renumber the figures with different figure numbers. (See also Chapter 8, Section J.)

__ Long view relationship between different parts not clear and unambiguous. 37 CFR 1.84(h)(2)(ii). Fig(s)_____

Probable cause of objection: A long object broken into several sections to fit on a single sheet is not arranged logically.

Remedy: Rearrange the parts so that they are arranged in sequence. (See Chapter 6, Section C3.)

Sectional views. 37 CFR 1.84(h)(3)

__ Hatching not indicated for sectional portions of an object. Fig(s)_____

Probable cause of objection: Sectioned or cutaway parts lack hatching.

Remedy: Provide hatching on the sectioned parts. (See Chapter 6, Sections C4 and C8.)

__ Hatching of regularly spaced oblique parallel lines not spaced sufficiently. Fig(s)_____

Probable cause of objection: Hatching lines are spaced too far apart or too close together.

Remedy: Space the hatching lines more appropriately, so that they are far enough apart to be distinguished from one another, but not too far apart so that they are too sparse. (See Chapter 6, Sections C4 and C8.)

__ Hatching not at substantial angle to surrounding axes or principal lines. Fig(s)_____

Probable cause of objection: Hatching lines are too close to vertical or horizontal, or are too parallel to edge (primary) lines so that they are difficult to distinguish.

Remedy: Angle the hatch lines so that they are about 45 degrees from horizontal, or at whatever angle necessary so that they may be easily distinguished from the edge lines. (See Chapter 6, Sections C4 and C8.)

__ Cross section not drawn same as view with parts in cross section with regularly spaced parallel oblique strokes. Fig(s)_____

Probable cause of objection: The sectional view does not show all the parts of the object that should be visible in cross section. For example, the figure only shows selected parts in cross section, when other parts that lie along the sectioning plane are not shown in cross section as they should be.

Remedy: Show all the parts that would be visible in cross section, as if the actual object is sliced in half. (See Chapter 6, Sections C4 and C8.)

__ Hatching of juxtaposed different elements not angled in a different way. Fig(s)_____

Probable cause of objection: Hatching on adjacent parts in a sectional view are at the same or very similar angles.

Remedy: Hatch adjacent parts at clearly different angles, preferably at opposite angles. (See Chapter 6, Sections C4 and C8.)

Alternate position. 37 CFR 1.84(h)(4)

__ A separate view required for a moved position. Fig(s)_____

Probable cause of objection: A part is shown in two different positions in solid lines; or one position in solid lines and another position in dashed lines, but the presence of the dashed part in the same figure is too confusing.

Remedy: Provide a separate figure to show the part in the moved position. (See Chapter 6, Section D.)

Modified forms. 37 CFR 1.84(h)(5)

__ **Modified forms of construction must be shown in separate views. Fig(s)_____**

Probable cause of objection: Different embodiments (versions) of the invention are depicted in the same figure.

Remedy: Provide separate figures for the different embodiments. (See Chapter 6, Section G.)

8. Arrangement of Views. 37 CFR 1.84(i)

__ **View placed upon another view or within outline of another. Fig(s)_____**

Probable cause of objection: Separate figures are placed too close together or even overlapping.

Remedy: Space the figures farther apart. (See Chapter 8, Section C1.)

__ **Words do not appear in a horizontal, left-to-right fashion when page is either upright or turned so that the top becomes the right side, except for graphs. Fig(s)_____**

Probable cause of objection: Some text is not oriented upright or horizontal with respect to the top of the sheet.

Remedy: Rearrange the text so that all of it is oriented upright and horizontal with respect to the top of the sheet, depending on whether the sheet is in portrait or landscape orientation. (See Chapter 8, Section A5.)

9. Scale. 37 CFR 1.84(k)

__ **Scale not large enough to show mechanism without crowding when drawing is reduced in size to two-thirds in reproduction. Fig(s)_____**

Probable cause of objection: The figure is too small to show details clearly.

Remedy: Enlarge the figure to show all of its details clearly. (See Chapter 8, Section J.)

__ **Indication such as "actual size" or "scale 1/2" not permitted. Fig(s)_____**

Probable cause of objection: Drawing scale indications such as "Actual Size," "1:2," "Scale 1/3," etc. are used.

Remedy: Delete all drawing scale indications. (See Chapter 8, Section N.)

__ **Elements of same view not in proportion to each other. Fig(s)_____**

Probable cause of objection: Some elements in the same figure are drawn out of proportion.

Remedy: Draw all elements of the same figure in proportion to each other. (See Chapter 8, Section J.)

10. Character of Lines, Numbers, & Letters. 37 CFR 1.84(l)

__ **Lines, numbers & letters not uniformly thick and well defined, clean, durable, and black (except for color drawings). Fig(s)_____**

Probable cause of objection: Lines or text are not sharp, uniformly thick, or black enough. This is the most common objection of all. Unacceptable lines are usually produced by improper pens, paper, drawing techniques, software, or printer.

For example, pencils, ball point pens, and roller ball pens usually do not produce acceptable lines, although some good quality roller pens may pass muster. Freehand lines, even if sharp, are usually not acceptable. Rough, highly porous paper will cause the ink to feather (seep out). Bitmap drawing programs usually produce very jagged lines. Dot matrix printers also produce jagged lines. Ink jet printers will not produce acceptable output if a paper not designed for ink jet printers is used; such papers tend to cause the ink to feather or seep out. Ink jet printers may, but do not always, produce acceptable output if special ink jet paper is used. Even laser printers may produce unacceptable lines, usually because the lines are too thin (at or less than 0.1 mm), or because the lines are too jagged, which is typical of 300 dpi printers.

Remedy: Redraw the specified figures with the proper pens and techniques; thicken the thinnest lines and reprint the drawings; reprint the drawings with a laser printer instead of an ink jet printer; or reprint the drawings with a 600 dpi printer. (See Chapter 8, Section H, and Chapters 2 and 3.)

11. Shading. 37 CFR 1.84(m)

__ **Shading used for other than shape of spherical, cylindrical, and conical elements of an object, or for flat parts. Fig(s)_____**

Probable cause of objection: Areas outside the object—for example—the background, are shaded.

Remedy: Delete shading from all areas other than on the object itself. (See Chapter 7, Section H.)

__ **Solid black shading areas not permitted. Fig(s)_____**

Probable cause of objection: Areas are filled with sold black.

Remedy: Delete the solid black areas, and leave an open outline. (See Chapter 7, Section I, and Chapter 8, Section N.)

12. Numbers, Letters, & Reference Characters. 37 CFR 1.84(p)

__ **Numbers and reference characters not plain and legible. 37 CFR 1.84(p)(1). Fig(s)_____**

Probable cause of objection: Text is poorly formed, either due to improper pens, techniques, software, or printer. As stated, pencils, ball point pens, and roller ball pens usually do not produce acceptable text. Freehand text is usually not acceptable, unless done extremely neatly and professionally. Fonts or writing with very thin lines (at or less than 0.1 mm) are usually not acceptable. Bitmap drawing programs usually produce very jagged text. Dot matrix printers also produce jagged text. Ink jet printers may, but do not always, produce acceptable text if special ink jet paper is used. All laser printers should produce acceptable text.

Remedy: Redraw the specified figures with the proper pens and techniques; use a thicker font or pen; print on special ink jet paper; or reprint the drawings with a laser printer instead of an ink jet printer. (See Chapters 2 and 3.)

__ **Numbers and reference characters used in conjunction with brackets, inverted commas, or enclosed within outlines. 37 CFR 1.84(p)(1). Fig(s)_____**

Probable cause of objection: Reference numbers or letters are surrounded by brackets, circles, boxes, etc.; or are written as primed numbers or letters—for example, 34'. Primed numbers or letters are not specifically prohibited, but are discouraged.

Remedy: Delete all brackets, circles, boxes, etc. from the reference numbers or letters, and replace the primed numbers or letters with different numbers or letters. (See Chapter 8, Sections D5 and N.)

__ **Numbers and reference characters not oriented in same direction as the view. 37 CFR 1.84(p)(1). Fig(s)_____**

Probable cause of objection: Some text is not oriented upright or horizontal with respect to the top of the sheet.

Remedy: Rearrange the text so that all of it is oriented upright and horizontal with respect to the top of the sheet, depending on whether the sheet is in portrait or landscape orientation. (See Chapter 8, Section A5.)

__ **English alphabet not used. 37 CFR 1.84(p)(2). Fig(s)_____**

Probable cause of objection: Non-English letters are used.

Remedy: Use English letters, or use numbers. (See Chapter 8, Sections D3 and D4.)

__ **Numbers, letters, and reference characters do not measure at least .32 cm. (1/8 inch) in height. 37 CFR 1.84(p)(3). Fig(s)_____**

Probable cause of objection: Text is too small.

Remedy: Make all text, including lower case letters, at least 1/8", 3.2 mm, or 14 points tall. (See Chapter 8, Sections D7 and D8.)

13. Lead Lines. 37 CFR 1.84(q)

__ **Lead lines cross each other. Fig(s)_____**

Probable cause of objection: Some lead lines cross each other.

Remedy: Reposition the lead lines, and also reference numbers if necessary, so that the lines do not cross each other. (See Chapter 8, Section E.)

__ **Lead lines missing. Fig(s)_____**

Probable cause of objection: One or more lead lines are missing for some reference numbers.

Remedy: Provide a lead line between each reference number or letter and its corresponding part in the drawings. (See Chapter 8, Section E.)

__ **Lead lines not as short as possible. Fig(s)_____**

Probable cause of objection: Lead lines are excessively long.

Remedy: Reposition the reference numbers reasonably close to the parts they are designating, to shorten the lead lines. (See Chapter 8, Section E.)

14. Numbering of Sheets of Drawings. 37 CFR 1.84(t)

__ **Number appears in top margin. Fig(s)_____**

Probable cause of objection: The sheet number intrudes into the top margin.

Remedy: Position the sheet number below the top margin by at least a few millimeters. (See Chapter 8, Section A6.)

__ **Number not larger than reference characters. Fig(s)_____**

Probable cause of objection: The sheet number is too small.

Remedy: Make the sheet number about 1/5", 5 mm, or 22 points in height. (See Chapter 8, Section A6.)

__ **Sheets not numbered consecutively, and in Arabic numerals, beginning with number 1. Sheets(s)_____**

Probable cause of objection: Drawing sheets are not numbered consecutively, are in non-Arabic numerals (something other than 1, 2, 3, etc.), or they do not start with 1.

Remedy: Renumber the sheets in consecutive Arabic numerals, starting with 1. (See Chapter 8, Section A6.)

15. Numbering of Views. 37 CFR 1.84(u)

__ **Views not numbered consecutively, and in Arabic numerals, beginning with number 1. Fig(s)_____**

Probable cause of objection: Figures are not numbered consecutively, are in non-Arabic numerals (something other than 1, 2, 3, etc.), or they do not start with 1.

Remedy: Renumber the figures in consecutive Arabic numerals, starting with 1. (See Chapter 8, Section C2.)

__ **View numbers not preceded by the abbreviation Fig. Fig(s)_____**

Probable cause of objection: Figure numbers do not include "Fig."

Remedy: Number figures as "Fig. 1," "Fig. 2," etc. (See Chapter 8, Section C2.)

__ **Single view contains a view number and the abbreviation Fig.**

Probable cause of objection: In an application with a single figure, the figure is designated with a figure number.

Remedy: Delete the figure number if the application include only one figure. (See Chapter 8, Section C2.)

__ **Numbers not larger than reference characters. Fig(s)_____**

Probable cause of objection: Figure numbers are too small.

Remedy: Make the figure numbers about 1/5", 5 mm, or 22 points in height. (See Chapter 8, Section C2.)

16. Corrections. 37 CFR 1.84(w)

__ **Corrections not durable and permanent. Fig(s)_____**

Probable cause of objection: Corrections were applied over a white correction fluid, and the ink is rubbing off, or the correction fluid is cracking or flaking off.

Remedy: Use a more durable ink on the correction fluid; submit a photocopy without any correction fluid on it; reprint the drawing; or redo the drawing. (See Chapter 8, Section M.)

17. Design Drawing. 37 CFR 1.152

__ **Surface shading shown not appropriate. Fig(s)_____**

Probable cause of objection: Inappropriate shading that does not properly depict the surface shape or contour, or the shading is of an inappropriate type.

Remedy: Redo the shading according to accepted techniques, and to properly depict the desired shape. (See Chapter 7, Section H.)

__ **Solid black shading not used for color contrast. Fig(s)_____**

Probable cause of objection: Solid black areas are used.

Remedy: Delete the solid back areas, and leave an open outline. (See Chapter 8, Section N.)

I. Do Not Add New Matter

Many defects cited by the objections and rejections, such as unclear description or drawings that lack sufficient detail, are seemingly curable by changing the description or drawings. However, after an application is filed, no new matter can be added to the same application. (New matter is any technical information not in the description, claims, or drawings as originally filed.) The objections and rejections cannot be overcome by substantively changing the information in the application, no matter how slightly; they must be overcome by relying on the information provided in the original application.

1. Changing Drawings Without Introducing New Matter

The new matter prohibition does not mean that the drawings cannot be changed. In a utility patent application, if a feature is not shown in the original drawings, but is specified in the original description and/or claims, it is not new matter, so that it may be properly added to the drawings if necessary.

> EXAMPLE 1: An application includes original drawings that show a portable CD player, but without a speaker. Fortunately, the written description states that a speaker is used. In this case, a speaker can be added to the drawings without violating the new matter prohibition.

Unless the element being added is very simple or readily understood by one skilled in the art—such as a hinge, a switch, a power supply, etc.—the original description must be detailed enough to support the changes to the drawings. However, the feature added to the drawings cannot have any more detail than the original description.

> EXAMPLE 2: The written description of a carpet cleaning method states that "a sprayer can be used" to apply a cleaning solution, but no

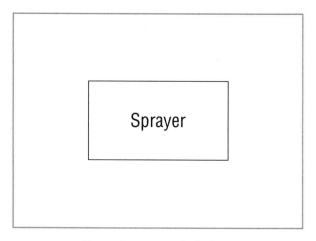

Illustration 9.2—Labeled Box

sprayer is shown in the original drawings. The drawings can be changed to show only the most basic, rudimentary sprayer. Alternatively, the sprayer may be represented with a labeled rectangular box to avoid any question of new matter, as in Illustration 9.2.

Generally, features cannot be added to design patent drawings, because there is no detailed description or claims to support them.

2. Adding New Matter by CIP

Although new matter cannot be added to the same application, it may be added to a *new* application, known as a "continuation-in-part" (CIP) application. (See *Patent It Yourself*, Chapter 14, for more on CIP applications.) However, the new matter will not get the benefit of the original filing date, so any relevant invention that predates the new matter may be used by the PTO to reject the CIP application.

> EXAMPLE: Maria Spinoza invented a chair on Jan. 1, 1996, and filed a patent application for it on the same day. On June 1, 1996, she added casters (wheels) to the legs of the chair; she then filed a CIP application on the same day to include the casters. However, on May

30, 1996, another inventor, Lucy Kant, created an identical chair with casters, and filed an application for it on that same day. Maria was unable to obtain a patent on a chair with casters because Lucy's chair with casters predated Maria's CIP application.

Also, under the 20-year-from-filing patent term, any patent based on a CIP application will expire 20 years from the filing date of the original (parent) application. The CIP application can lengthen the time it takes to get a patent from several months to a couple of years, so the patent term is effectively reduced by the same amount. For these reasons, CIP applications should be avoided by ensuring that the original drawings, as well as the original written description, include enough information to enable one skilled in the pertinent field to make and use the invention.

3. Removing Features

Removing a feature may be considered as introducing new matter if its removal changes the invention. For design inventions, *nothing may be removed*, because the appearance of a design invention cannot be changed after filing, not even slightly. However, if the feature was described in the original application as not being part of the invention, or if the feature was shown in dashed lines (which are used to illustrate features that are not part of the invention; see Chapter 7, Section G) in the original drawings, it may be removed without introducing new matter—that is, altering the design.

EXAMPLE 1: A utility application for a lamp describes and shows a switch for turning the lamp on and off at will. The switch cannot be deleted from the application, because its removal will substantively change the invention—that is, the lamp will turn on and stay on all the time when it is plugged in—and introduce new matter.

EXAMPLE 2: A design application for a clock shows a plain clock face with a single round dot at the 12 o'clock position. The dot cannot be deleted from the drawings without introducing new matter. The application also shows a power cord. If the power cord is shown in solid lines, it cannot be deleted from the drawings without introducing new matter. However, if the power cord is shown in solid lines, but the description for the figures states that the power cord is not part of the invention, it may be deleted from the drawings without introducing new matter. Also, if the power cord is shown in dashed lines, even if there is no statement indicating that it is not part of the invention, it may be deleted from the drawings without introducing new matter.

J. Correcting the Drawings

The PTO will not correct the drawings for you or return them to you. New drawings or corrected copies must be submitted. Therefore, if you made the drawings by hand, you should keep the original in case corrections are required. If you made them with a computer, keep the computer file so you may make changes later. (It's always a good idea to keep back-ups.) Be careful not to introduce new matter; refer to Section I, above, for details.

1. Correcting Ink Drawings

If you made the drawings by hand, you must correct the original and submit a good quality photocopy free of copier specks. Drawings made on vellum, which is a durable and substantially non-porous paper, may be erased with a suitable eraser (see Chapter 2, Section A) without damaging the paper. The elements to be deleted must be completely erased, so that no trace of such elements appear on a photocopy.

Drawings made on porous paper cannot be erased without damaging or fraying the paper. Ink applied on the frayed areas will feather, so the drawings will again be objected to. Therefore, it is best to use white correction fluid on porous paper. Correction fluid does not provide an ideal surface for ink; so if large areas are covered, the lines applied over it will probably be poor, which may again be objected to. Redoing the entire sheet may be preferable in such a situation.

If the figures on a sheet must be repositioned, they may be cut out, taped, or glued onto a new sheet, and photocopied. Make sure that the edges of the pasted pieces do not appear on the photocopy. If they do, try taping or gluing down the offending edge, or covering the unwanted lines of the photocopy with white correction fluid.

2. Correcting CAD Drawings

Correcting CAD drawings is easy. Just open the drawing file, make the necessary changes on screen, and print a new copy. This is one of the greatest advantages of making drawings with a computer.

3. Adding an Element to a Drawing

If an element is specified in the description but not shown in the original drawings, it may optionally be added to the drawings. If the element is specified in the claims but not shown in the original drawings, it must be added to the drawings, or the element must be deleted from the claims. Such a requirement is typically stated in an Office Action.

Prior to making any changes to the drawings (other than to make minor corrections, or corrections in response to the Notice of Draftsperson's Drawing Review), you must first obtain permission from the examiner. To obtain permission, file a copy of the drawing sheet to be changed, with the element being added shown in red, along with a copy of the Request for Entry of Drawing Amendment which is in the Appendix.

You can make a photocopy of the original drawing, and draw the element in red ink. If you are drawing with a computer, you may print the drawing with the new element in black, and trace over or circle it with red ink. It is expedient to concurrently submit an extra copy of the drawing sheet with the new element properly executed in black lines. If you do this, add the following sentence to Request for Entry of Drawing Amendment: "A formal copy of the drawing with the proposed changes is also enclosed."

4. Making Voluntary Changes

If you think changes to the substance of the figures are proper and necessary—such as adding or changing elements—you must, as stated, request approval for such changes by the examiner. Submit a new drawing with the changes in red, and explain why they are necessary on a separate sheet. Use the same procedure discussed in Section 3, above, for adding new elements to a drawing.

To obtain permission to delete elements from the drawings, make a copy of the drawing sheet, use red ink to cross out the elements or figures to be deleted, and submit the red-inked copy with a Request for Entry of Drawing Amendment (a tear-out is in the Appendix). As stated, it is expedient to concurrently submit an extra copy of the drawing sheet without the deleted element. If you do this, add the following sentence to the Request for Entry of Drawing Amendment: "A formal copy of the drawing with the proposed changes is also enclosed."

K. Filing Corrected Drawings

Subject to approval by the examiner, voluntary changes may be filed any time before the application is allowed. Corrections required by the examiner or the draftsperson may be filed with the written response (see *Patent It Yourself,* Chapter 13, for details on Office Action responses), or you may choose to file them after allowance—that is, after the application is approved. If they are filed after allowance, they must be filed within three months from the mailing date of the Notice of Allowance (a notice indicating that the application is allowable). They should be filed as soon as possible after allowance, because if the drawings are again objected to, another set of corrected drawings must be filed within the same three-month period.

The time for filing corrected drawings may be extended for up to three additional months by filing a petition for extension of time and paying the appropriate petition fee. (See *Patent It Yourself,* Chapter 13, for details on extensions.) However, the time for paying the issue fee cannot be

extended. We recommend that you do not put off making the corrections and avoid paying the petition fees, because you can better use your money to promote your invention.

L. Summary

Congratulations on finishing the book! As with any endeavor that requires a new skill, you may not be able to turn out patent drawings immediately. There are learning curves associated with ink drafting and computer drafting. Be patient in your efforts to get the equipment together, and in your initial attempts to use them. Reread the relevant sections of the book as necessary. In the end, you will find that your work will pay off in savings, and in having the satisfaction of being able to create professional-looking patent drawings yourself. ∎

Tear-out Forms

Petition for Submitting Black-and-White Photographs

Petition for Submitting Color Photographs or Drawings

Request for Entry of Drawing Amendment

App. No: _____

Filing Date: _____

Applicant: _____

App. Title: _____

Examiner: _____

Art Unit: _____

Petition for Submitting Black-and-White Photographs

Assistant Commissioner for Patents
Washington, DC 20231

Sirs:

Applicant hereby respectfully petitions that the black-and-white photographs ☐ filed herewith ☐ already filed be accepted as formal drawings. The petition fee is filed herewith.

Sole/First Applicant: _____

_____   _____
Sole/First Applicant Signature   Date

Joint/Second Applicant: _____

_____   _____
Joint/Second Applicant Signature   Date

Sole/Third Applicant: _____

_____   _____
Sole/Third Applicant Signature   Date

Sole/Fourth Applicant: _____

_____   _____
Sole/Fourth Applicant Signature   Date

App. No: _____

Filing Date: _____

Applicant: _____

App. Title: _____

Examiner: _____

Art Unit: _____

Petition for Submitting Color Photographs or Drawings

Assistant Commissioner for Patents
Washington, DC 20231

Sirs:

Applicant hereby respectfully petitions that the color photographs ☐ filed herewith ☐ already filed be accepted as formal drawings. The petition fee is filed herewith.

These color photographs or drawings are necessary because _____

_____

_____

_____

_____

_____

_____

_____

_____

Sole/First Applicant: _____

_____   _____
Sole/First Applicant Signature   Date

Joint/Second Applicant: _____

_____   _____
Joint/Second Applicant Signature   Date

Sole/Third Applicant: _____

_____   _____
Sole/Third Applicant Signature   Date

Sole/Fourth Applicant: _____

_____   _____
Sole/Fourth Applicant Signature   Date

App. No: _____

Filing Date:_____

Applicant: _____

App. Title: _____

Examiner: _____

Art Unit: _____

Request for Entry of Drawing Amendment

Assistant Commissioner for Patents
Washington, DC 20231

Sirs:

Applicant respectfully requests entry of the drawing amendment indicated in red on the drawing submitted herewith.

Respectfully,

_____
Applicant

Certificate Of Mailing

I hereby certify that this paper will be deposited on the date below with the United States Postal Service as First Class Mail, postage prepaid, in an envelope addressed to "Assistant Commissioner for Patents, Washington, DC 20231"

_____   _____
Person Depositing Paper   Date

Index

CATALOG
...more from Nolo Press

	EDITION	PRICE	CODE

BUSINESS

	EDITION	PRICE	CODE
Business Plans to Game Plans	1st	$29.95	GAME
The California Nonprofit Corporation Handbook	7th	$29.95	NON
The California Professional Corporation Handbook	5th	$34.95	PROF
The Employer's Legal Handbook	2nd	$29.95	EMPL
Form Your Own Limited Liability Company	1st	$24.95	LIAB
Helping Employees Achieve Retirement Security	1st	$16.95	HEAR
Hiring Independent Contractors: The Employer's Legal Guide, (Book w/Disk—PC)	2nd	$29.95	HICI
How to Finance a Growing Business	4th	$24.95	GROW
How to Form a CA Nonprofit Corp.—w/Corp. Records Binder & PC Disk	1st	$49.95	CNP
How to Form a Nonprofit Corp., Book w/Disk (PC)—National Edition	3rd	$39.95	NNP
How to Form Your Own Calif. Corp.—w/Corp. Records Binder & Disk—PC	1st	$39.95	CACI
How to Form Your Own California Corporation	8th	$29.95	CCOR
How to Form Your Own Florida Corporation, (Book w/Disk—PC)	3rd	$39.95	FLCO
How to Form Your Own New York Corporation, (Book w/Disk—PC)	3rd	$39.95	NYCO
How to Form Your Own Texas Corporation, (Book w/Disk—PC)	4th	$39.95	TCOR
How to Handle Your Workers' Compensation Claim (California Edition)	1st	$29.95	WORK
How to Market a Product for Under $500	1st	$29.95	UN500
How to Mediate Your Dispute	1st	$18.95	MEDI
How to Write a Business Plan	4th	$21.95	SBS
The Independent Paralegal's Handbook	4th	$29.95	PARA
Insuring the Bottom Line	1st	$29.95	BOTT
Legal Guide for Starting & Running a Small Business, Vol. 1	3rd	$24.95	RUNS
Make Up Your Mind: Entrepreneurs Talk About Decision Making	1st	$19.95	MIND

Book with disk

	EDITION	PRICE	CODE
Managing Generation X: How to Bring Out the Best in Young Talent	1st	$19.95	MANX
Marketing Without Advertising	2nd	$19.00	MWAD
Mastering Diversity: Managing for Success Under ADA and Other Anti-Discrimination Laws	1st	$29.95	MAST
OSHA in the Real World: (Book w/Disk—PC)	1st	$29.95	OSHA
Pay For Results	1st	$29.95	PAY
The Partnership Book: How to Write a Partnership Agreement, (Book w/Disk—PC)	5th	$34.95	PART
Rightful Termination	1st	$29.95	RITE
Sexual Harassment on the Job	2nd	$18.95	HARS
Taking Care of Your Corporation, Vol. 1, (Book w/Disk—PC)	1st	$26.95	CORK
Taking Care of Your Corporation, Vol. 2, (Book w/Disk—PC)	1st	$39.95	CORK2
Tax Savvy for Small Business	2nd	$26.95	SAVVY
Trademark: How to Name Your Business & Product	2nd	$29.95	TRD
Workers' Comp for Employers	2nd	$29.95	CNTRL
Your Rights in the Workplace	3rd	$19.95	YRW

CONSUMER

	EDITION	PRICE	CODE
Fed Up With the Legal System: What's Wrong & How to Fix It	2nd	$9.95	LEG
Glossary of Insurance Terms	6th	$14.95	GLINT
How to Insure Your Car	1st	$12.95	INCAR
How to Insure Your Home	1st	$12.95	INTRO
How to Insure Your Life	1st	$12.95	INLIF
How to Win Your Personal Injury Claim	2nd	$24.95	PICL
Nolo's Everyday Law Book	1st	$21.95	EVL
Nolo's Pocket Guide to California Law	5th	$11.95	CLAW
The Over 50 Insurance Survival Guide	1st	$16.95	OVER50
Trouble-Free Travel...And What to Do When Things Go Wrong	1st	$14.95	TRAV
True Odds: How Risk Affects Your Everyday Life	1st	$19.95	TROD
What Do You Mean It's Not Covered?	1st	$19.95	COVER

ESTATE PLANNING & PROBATE

	EDITION	PRICE	CODE
8 Ways to Avoid Probate (Quick & Legal Series)	1st	$15.95	PRO8
How to Probate an Estate (California Edition)	9th	$34.95	PAE
Make Your Own Living Trust	2nd	$21.95	LITR
Nolo's Will Book, (Book w/Disk—PC)	3rd	$24.95	SWIL
Plan Your Estate	3rd	$24.95	NEST
The Quick and Legal Will Book	1st	$15.95	QUIC
Nolo's Law Form Kit: Wills	1st	$14.95	KWL

Book with disk

	EDITION	PRICE	CODE

FAMILY MATTERS

	EDITION	PRICE	CODE
A Legal Guide for Lesbian and Gay Couples	9th	$24.95	LG
California Marriage Law	12th	$19.95	MARR
Child Custody: Building Parenting Agreements that Work	2nd	$24.95	CUST
Divorce & Money: How to Make the Best Financial Decisions During Divorce	3rd	$26.95	DIMO
Get A Life: You Don't Need a Million to Retire Well	1st	$18.95	LIFE
The Guardianship Book (California Edition)	2nd	$24.95	GB
How to Adopt Your Stepchild in California	4th	$22.95	ADOP
How to Do Your Own Divorce in California	21st	$24.95	CDIV
How to Do Your Own Divorce in Texas	6th	$19.95	TDIV
How to Raise or Lower Child Support in California	3rd	$18.95	CHLD
The Living Together Kit	8th	$24.95	LTK
Nolo's Law Form Kit: Hiring Childcare & Household Help	1st	$14.95	KCHD
Nolo's Pocket Guide to Family Law	4th	$14.95	FLD
Practical Divorce Solutions	1st	$14.95	PDS
Smart Ways to Save Money During and After Divorce	1st	$14.95	SAVMO

GOING TO COURT

	EDITION	PRICE	CODE
Collect Your Court Judgment (California Edition)	3rd	$24.95	JUDG
The Criminal Records Book (California Edition)	5th	$21.95	CRIM
How to Sue For Up to 25,000...and Win!	2nd	$29.95	MUNI
Everybody's Guide to Small Claims Court in California	12th	$18.95	CSCC
Everybody's Guide to Small Claims Court (National Edition)	6th	$18.95	NSCC
Fight Your Ticket ... and Win! (California Edition)	6th	$19.95	FYT
How to Change Your Name (California Edition)	6th	$24.95	NAME
Mad at Your Lawyer	1st	$21.95	MAD
Represent Yourself in Court: How to Prepare & Try a Winning Case	1st	$29.95	RYC
Taming the Lawyers	1st	$19.95	TAME

HOMEOWNERS, LANDLORDS & TENANTS

	EDITION	PRICE	CODE
The Deeds Book (California Edition)	4th	$16.95	DEED
Dog Law	2nd	$12.95	DOG
⌨ Every Landlord's Legal Guide (National Edition)	1st	$34.95	ELLI
For Sale by Owner (California Edition)	2nd	$24.95	FSBO
Homestead Your House (California Edition)	8th	$9.95	HOME
How to Buy a House in California	4th	$24.95	BHCA
The Landlord's Law Book, Vol. 1: Rights & Responsibilities (California Edition)	5th	$34.95	LBRT
The Landlord's Law Book, Vol. 2: Evictions (California Edition)	6th	$34.95	LBEV

⌨ Book with disk

CALL 800-992-6656 OR USE THE ORDER FORM IN THE BACK OF THE BOOK

	EDITION	PRICE	CODE

Leases & Rental Agreements (Quick & Legal Series) .... 1st | $18.95 | LEAR
Neighbor Law: Fences, Trees, Boundaries & Noise .... 2nd | $16.95 | NEI
Safe Homes, Safe Neighborhoods: Stopping Crime Where You Live .... 1st | $14.95 | SAFE
Tenants' Rights (California Edition) .... 13th | $19.95 | CTEN

HUMOR

29 Reasons Not to Go to Law School .... 1st | $9.95 | 29R
Poetic Justice .... 1st | $9.95 | PJ

IMMIGRATION

How to Become a United States Citizen .... 5th | $14.95 | CIT
How to Get a Green Card: Legal Ways to Stay in the U.S.A. .... 2nd | $24.95 | GRN
U.S. Immigration Made Easy .... 5th | $39.95 | IMEZ

MONEY MATTERS

Building Your Nest Egg With Your 401(k) .... 1st | $16.95 | EGG
Chapter 13 Bankruptcy: Repay Your Debts .... 2nd | $29.95 | CH13
Credit Repair (Quick & Legal Series) .... 1st | $15.95 | CREP
How to File for Bankruptcy .... 6th | $26.95 | HFB
Money Troubles: Legal Strategies to Cope With Your Debts .... 4th | $19.95 | MT
Nolo's Law Form Kit: Personal Bankruptcy .... 1st | $14.95 | KBNK
Simple Contracts for Personal Use .... 2nd | $16.95 | CONT
Stand Up to the IRS .... 3rd | $24.95 | SIRS
The Under 40 Financial Planning Guide .... 1st | $19.95 | UN40

PATENTS AND COPYRIGHTS

The Copyright Handbook: How to Protect and Use Written Works .... 3rd | $24.95 | COHA
Copyright Your Software .... 1st | $39.95 | CYS
Patent, Copyright & Trademark: A Desk Reference to Intellectual Property Law .... 1st | $24.95 | PCTM
Patent It Yourself .... 5th | $44.95 | PAT
⌨ Software Development: A Legal Guide (Book with disk—PC) .... 1st | $44.95 | SFT
The Inventor's Notebook .... 2nd | $19.95 | INOT

RESEARCH & REFERENCE

◎ Law on the Net, (Book w/CD-ROM—Windows/Macintosh) .... 2nd | $39.95 | LAWN
Legal Research: How to Find & Understand the Law .... 4th | $19.95 | LRES
Legal Research Made Easy (Video) .... 1st | $89.95 | LRME

◎ Book with CD-ROM
⌨ Book with disk

CALL 800-992-6656 OR USE THE ORDER FORM IN THE BACK OF THE BOOK

	EDITION	PRICE	CODE

SENIORS

Beat the Nursing Home Trap	2nd	$18.95	ELD
Social Security, Medicare & Pensions	6th	$19.95	SOA
The Conservatorship Book (California Edition)	2nd	$29.95	CNSV

SOFTWARE

California Incorporator 2.0—DOS	2.0	$47.97	INCI2
Living Trust Maker 2.0—Macintosh	2.0	$47.97	LTM2
Living Trust Maker 2.0—Windows	2.0	$47.97	LTWI2
Small Business Legal Pro—Macintosh	2.0	$25.97	SBM2
Small Business Legal Pro—Windows	2.0	$25.97	SBW2
Small Business Legal Pro Deluxe CD—Windows/Macintosh CD-ROM	2.0	$35.97	SBCD
Nolo's Partnership Maker 1.0—DOS	1.0	$47.97	PAGI1
Personal RecordKeeper 4.0—Macintosh	4.0	$29.97	RKM4
Personal RecordKeeper 4.0—Windows	4.0	$29.97	RKP4
Patent It Yourself 1.0—Windows	1.0	$149.97	PYW1
WillMaker 6.0—Macintosh	6.0	$29.97	WM6
WillMaker 6.0—Windows	6.0	$29.97	WIW6

⌐ Book with disk

CALL 800-992-6656 OR USE THE ORDER FORM IN THE BACK OF THE BOOK

ORDER FORM

Code	Quantity	Title	Unit price	Total

Subtotal	
California residents add Sales Tax	
Basic Shipping ($6.00 for 1 item; $7.00 for 2 or more)	
UPS RUSH delivery $7.50–any size order*	
TOTAL	

Name

Address

(UPS to street address, Priority Mail to P.O. boxes)

* Delivered in 3 business days from receipt of order. S.F. Bay Area use regular shipping.

FOR FASTER SERVICE, USE YOUR CREDIT CARD AND OUR TOLL-FREE NUMBERS

Order 24 hours a day	1-800-992-6656
Fax your order	1-800-645-0895
e-mail	cs@nolo.com
General Information	1-510-549-1976
Customer Service	1-800-728-3555, Mon.-Fri. 9am-5pm, PST

METHOD OF PAYMENT

☐ Check enclosed

☐ VISA   ☐ MasterCard   ☐ Discover Card   ☐ American Express

Account #   Expiration Date

Authorizing Signature

Daytime Phone

PRICES SUBJECT TO CHANGE.

VISIT OUR OUTLET STORES!

You'll find our complete line of books and software, all at a discount.

BERKELEY
950 Parker Street
Berkeley, CA 94710
1-510-704-2248

SAN JOSE
111 N. Market Street, #115
San Jose, CA 95113
1-408-271-7240

VISIT US ONLINE!

on the INTERNET — www.nolo.com

NOLO PRESS 950 PARKER ST., BERKELEY, CA 94710

Patent, Copyright and Trademark

A Desk Reference to Intellectual Property Law

by Attorney Stephen Elias

$24.95 • PCTM

This invaluable bestseller provides a plain-English, easy-to-understand overview of patent, copyright, trademark and trade secret law. Clear concise definitions, contextual examples, and sample forms and agreements demystify one of the most complex and confusing areas of the law.

The Copyright Handbook

How to Protect and Use Written Works

by Attorney Stephen Fishman

$24.95 • COHA

This bestselling book provides step-by-step instructions and all the forms necessary to protect all kinds of written expression under U.S. and international copyright law, including works produced on CD-ROM or on disk, computer databases and electronic mail.

License Your Invention

Take Your Idea to Market with a Solid Legal Agreement

by Attorney Richard Stim

$39.95 • LYI

License Your Invention tells inventors everything they need to know to enter into a good written agreement with the manufacturer, marketer or distributor who will handle the details of merchandising the invention. It shows step-by-step how to draft a license that will be fair to all parties. Sample tear-out agreement included, and all forms are provided on disk.

Software Development

A Legal Guide

by Attorney Stephen Fishman

$44.95 • SOFT

This book is an invaluable guide to the tangle of complex legal rules regulating software development and protection. It includes step-by-step instructions and all the forms needed to register a software copyright with the U.S. Copyright Office, along with practical, thorough and comprehensive advice about copyright law. All contracts, agreements and legal forms are provided on disk.

Trademark

How to Name Your Business and Product

by Attorneys Kate McGrath & Stephen Elias with Trademark Attorney Sarah Shena

$29.95 • TRD

Essential for all small business owners, this book shows them how to choose, use and protect the names and symbols that identify their services or products. The newly revised third edition contains all the official forms and instructions necessary to register a federal trademark or servicemark with the U.S. Patent and Trademark Office.